THE
UNIVERSE
IN BITE-SIZED
CHUNKS

THE
UNIVERSE
IN BITE-SIZED
CHUNKS

COLIN STUART

Michael O'Mara Books Limited

For Mum and Dad.
Thank you for always encouraging
me to reach for the stars

First published in Great Britain in 2018
by Michael O'Mara Books Limited
9 Lion Yard
Tremadoc Road
London SW4 7NQ

A CIP catalogue record for this book is available from the British Library.

Papers used by Michael O'Mara Books Limited are natural,
recyclable products made from wood grown in sustainable forests.
The manufacturing processes conform to the environmental
regulations of the country of origin.

ISBN: 978-1-78243-864-9 in hardback print format
ISBN: 978-1-78243-866-3 in ebook format

1 2 3 4 5 6 7 8 9 10

Designed and typeset by Ed Pickford
Illustrations by Greg Stevenson

Printed and bound by CPI Group (UK) Ltd, Croydon, CR0 4YY

www.mombooks.com

Contents

Picture Credits

Introduction

'I have loved the stars too fondly to be fearful of the night.'

The Old Astronomer (To His Pupil), Sarah Williams (1868)

I've been captivated by the night sky for as long as I can remember. It was the first time I ever fell in love. As children we are told wonderful stories of goblins, dragons and witches, but the universe has always been more magical to me than any fairy tale.

Generations of astronomers have pulled back the curtain from the cosmos and revealed its innermost secrets. What they've found is nothing short of incredible. Countless planets dance around an endless expanse of stars. Gravity twists and curls space until time itself grinds to a halt. We can follow atoms on a journey all the way from the heart of stars to your skin and bones. We've sent machines to every planet in the solar system and left our footprints in the lunar dust.

The sheer scale of such a universe can be intimidating. I've spent the last ten years writing and speaking about astronomy and it still makes me feel small. A lot of people are put off because they assume learning about it must be difficult. But it doesn't need to be. The aim of this book is to break the vastness of space down into digestible pieces that are easy to understand. There's no maths or jargon here, just simple explanations of the universe's most fascinating features.

I've included as much about what we don't know as what we do. Answering one question throws up many others. We still don't understand what most of our universe is made of or whether we share space with any other life forms. Astronomers are still trying to figure out if our universe is the only one and exactly how space and time got started. These are some of the most fundamental questions it is possible to ask.

The book is organized in order of increasing distance from Earth, starting with our earliest astronomical discoveries before heading out into the wider solar system and then to the galaxy and universe(s) beyond. Our travels will cover 93 billion light years of space and nearly 14 billion years of time. I have carefully selected our itinerary so that you can hold the entire universe in your hand and discover what interests you most along the way.

So join me on a journey across the cosmos. I hope you will fall in love with the night sky, too.

CHAPTER 1

Early Astronomy

Marking the passage of time

Long before the sky was a place of planets, galaxies and black holes it was the realm of gods and omens. A crack of thunder could signal Almighty displeasure; a passing comet was an ominous harbinger of doom. At least that's how many of our ancestors saw it.

But the sky's most important role was as a natural timepiece. In the aeons before clocks, computers and smartphones, our predecessors noticed that the sky ticked out its own natural rhythm. The sun would come and go over a period they came to know as a day. They gathered together seven of these days to form a week, with each day named after one of the seven celestial objects they saw behaving differently to the stars (see page 18).

The moon changed its appearance, waxing and waning through phases, growing from a tiny crescent to a dazzling full moon and back again. One cycle of this shape-shifting took almost thirty days, a period they called a 'moonth'. The relentless morphing of language over time has seen us lose a letter. The sun, too, executes a much longer cycle. Rising each morning towards the east, and setting towards the west in the evening, it reaches the peak of its daily climb at midday. Yet its height above the ground at noon is not always the same. Watch over many months and you'll see the sun trace out a figure-of-eight shape in the sky called an 'analemma'. In the time it takes to complete this particular cycle the sun rises 365 times. The ancients called this cycle a year. This period was divided into four seasons, each with its own distinct weather trends. Winter, spring, summer and autumn were seen to repeat in the same time as the analemma took to complete.

The sun appears to trace out a 'figure of eight' in the sky over the course of a year. Astronomers call it an analemma.

By 10,000 years ago we were building massive clocks to keep up with the sky's natural rhythms. In 2004, a team of archaeologists discovered an ancient Stone Age site in Scotland dating from this time. By 2013 they had realized why it was built. The architects of the site dug twelve pits along an arc 50 metres long – one for each of the twelve complete lunar cycles which normally fit into a year (occasionally there can be thirteen full moons in a year if the first falls in early January). Five thousand years later, stonemasons began work on the mighty circle of Stonehenge on Salisbury Plain, in England. Standing inside the circle, you see the sun rise directly above one particular stone – the heel stone – on the day when it reaches the top of the analemma (the summer solstice).

Today we go about our hectic, modern lives in the digital age largely unaware of the rhythm of the sky. But for ancient civilizations it was the only way of measuring time, and their extensive studies of the movements of the sun and the stars form the basis of how we organize our lives today.

Discovering the shape of the Earth

Don't believe anyone who tells you that the best minds of the Middle Ages believed the world was flat – we knew that it wasn't more than 2,000 years ago. The man we have to thank for that knowledge is Ancient Greek mathematician Eratosthenes, and he figured it out without ever leaving Egypt.

He noted that in the Egyptian city of Syene the sun was directly overhead at noon on the summer solstice. His genius was to make a measurement of the sun at exactly the same time on a subsequent summer solstice in the city of Alexandria, some 800 kilometres away. By placing a stake in the ground, and looking at its shadow, he could see that the sun struck his stake not from overhead but at an angle of seven degrees. The reason for this difference is that the Earth's surface is curved, meaning the sun's light strikes each city at a different angle.

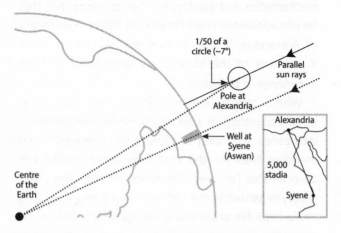

Eratosthenes worked out the size of the Earth by looking at the angle of shadows from different locations in Egypt.

He went one step further. If a distance of 800 kilometres causes a seven-degree difference, he could scale that up to see the distance represented by a full 360 degrees. That gives the circumference of the Earth as just over 41,000 kilometres (he did his calculations using an ancient unit of distance called the 'stadion', so his answer was actually approximately 250,000 'stadia'). He was within 10–15 per cent of our modern value for the Earth's size. So not only did the Ancient Greeks know the Earth was round, they also had a pretty good idea about how big it was, too.

ERATOSTHENES (256–194 BCE)

Eratosthenes was one of the original polymaths. Along with his work on the circumference of the Earth, he made important contributions to geography, music, mathematics and poetry. He was so respected that he was appointed chief librarian at the famous library of Alexandria. It later burned down, but at its peak it was one of the biggest repositories for ancient knowledge in the world.

With access to many important maps and scrolls, he put together an atlas of the world and divided it into zones according to climate. He drew grids and meridian lines for the first time and provided the co-ordinates for over 400 cities. For this work he's widely regarded as the father of geography.

Perhaps his second-biggest achievement was to invent the sieve of Eratosthenes – a way of identifying prime numbers by filtering out all those numbers whose repeating behaviour means they cannot be prime (a prime number can only be divided by two numbers – itself and one).

In recognition of his important work there is a crater on the moon named after him.

It is possible that humans knew about the shape of the Earth, if not its size, even before the time of Eratosthenes. During a partial lunar eclipse the Earth's shadow is cast upon the surface of the moon (see page 10). This shadow is very obviously curved. It has been speculated that a Chinese book called *Zhou-Shu* records the occurrence of a lunar eclipse in the twelfth century BCE. *The Clouds* – a Greek play by Aristophanes – certainly records a lunar eclipse from 421 BCE. If either civilization understood that what they were seeing was caused by the Earth preventing sunlight reaching the moon, then they'd have realized the Earth wasn't flat. And it is to eclipses that we turn next.

Solar eclipses

An eclipse is simply an event in the sky where something that is normally visible is blocked from view. Eclipses come in two main forms: solar and lunar. During a solar eclipse the sun is blocked by the moon; during a lunar eclipse the Earth prevents most light reaching the moon.

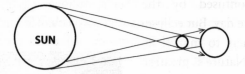

We see a solar eclipse when the moon blocks our view of the sun.

Humans have watched, wondered and worried at solar eclipses in particular for thousands of years. It is said that during the reign of the Chinese king Zhong Kang, he cut off the heads of his two court astronomers after they failed to predict a solar eclipse. That was 4,000 years ago. Before our modern understanding of the event, solar eclipses were often seen as bad omens – the gods showcasing their displeasure at the sins of mankind.

The most spectacular form of solar eclipse is a total one – when the moon entirely covers the solar disc. They are rare events for any particular location, but a total solar eclipse happens somewhere on Earth every eighteen months or so. The moon's swift passage across the sky means that the spectacle can never last more than seven minutes and thirty-two seconds. Arguably the most beautiful part of a solar eclipse is called Baily's beads, named after a nineteenth-century English astronomer. Just before and after the moment of totality, the last and first gasps of sunlight are able to make it to us only through craters on the very edge of the lunar surface. This gives rise to a stunning diamond ring effect.

During totality, the sky noticeably darkens and the temperature drops. Birds that were happily singing fall silent, confused by the sun's sudden disappearance during the day. But eclipses aren't just a chance for amateur skywatchers to marvel at one of Nature's greatest spectacles – they're an invaluable chance for astronomers to learn more about the cosmos. As we'll see, some of our landmark breakthroughs in understanding the universe have come on the back of observing total solar eclipses (see page 49).

The diamond ring effect known as Baily's beads.

Not all solar eclipses are total, however. Often the moon only covers some of the solar disc. During these partial solar eclipses it looks as if the sun has had a big bite taken out of it. The moon's distance from the Earth varies slightly, and so sometimes it's too far from us, and appears too small, to block out the sun entirely. We call these eclipses annular after the Latin word *annulus*, which means 'little ring'.

It's worth noting that we live in an exceptional time for solar eclipses. That's because millions of years ago the moon was much closer to the Earth (see page 83) and would have regularly blocked out the sun entirely, but without the grand spectacle of Baily's beads. In the future, as the moon recedes even further from us, it will eventually appear too small to ever gift us total solar eclipses. Our distant descendants will have to make do with only partial and annular eclipses.

Lunar eclipses

We only see the moon because it reflects sunlight. But during a total lunar eclipse all direct light from the sun is blocked by the Earth. The moon drifts into the Earth's shadow – or *umbra*. If the moon only moves through a part of Earth's shadow we get a partial or *penumbral* lunar eclipse.

While all *direct* sunlight is prevented from reaching the moon during totality, some indirect light still makes it to the lunar surface. That's thanks to the Earth's atmosphere bending – or *refracting* – a small amount of sunlight around our planet. White light is really a mix of the seven colours of the rainbow (see page 32) and our atmosphere bends the red light towards the moon – the rest is scattered off into space. That's why the moon turns various shades of copper, orange or red during a total lunar eclipse. Airborne volcanic ash intensifies the effect and renders the moon a deeper blood red. Without Earth's atmosphere, however, the full moon would appear to take leave temporarily from the sky.

Unlike solar eclipses, which are relatively rare and short-lived, lunar eclipses are reasonably frequent and last longer. It is much easier for a large object like the Earth to block light from reaching a small object like the moon than it is for the moon to blot out an enormous object like the sun. Totality can last up to 100 minutes and be seen by most people on the night side of the Earth.

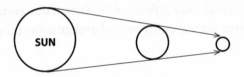

We see a lunar eclipse when the moon disappears
into Earth's shadow.

Humans have been observing lunar eclipses for millennia. Ancient Sumerian clay tablets dating from 2094 BCE record a lunar eclipse along with predictions of impending doom – superstition and eclipses often went hand in hand. The most notorious lunar eclipse occurred in 1504 CE just after Christopher Columbus had discovered the New World. The Italian explorer and his crew were holed up in Jamaica, forced to make repairs to his fleet of ships as worms ate away at the wooden hulls.

At first the locals were accommodating, but the visitors began to outstay their welcome and aggravated the natives by plundering their food. After six months the local chief cut off all supplies. Desperate, Columbus thought on his feet. All ships at the time carried almanacs – catalogues of star positions and astronomical events to help with navigation. He could see that a lunar eclipse was due for 29 February. In an act of pure cunning, Columbus told the chief that he was in contact with God and that His heavenly displeasure with the explorer's treatment would be shown by turning the moon blood red. When the eclipse duly came the following evening the locals were suddenly more cooperative.

According to a report by Columbus's son: 'With great howling and lamentation they came running from every direction to the ships, laden with provisions, praying the Admiral to intercede by all means with God on their behalf.' Such is the power of knowing how the universe really works and the danger of superstition.

The constellations

A long with the moon, the night sky is dominated by the stars. On a clear night thousands are visible and over millennia many independent civilizations have played giant games of dot-to-dot, linking them up in their imaginations to form groups known as constellations. Often they're completely arbitrary, with the stars in each pattern having little to do with one another save for their apparent proximity in our sky. Many are far from realistic. Take the constellation known as Canis Minor – The Little Dog. It is made from just two stars joined together by a single line. Hardly dog-like – there aren't even any legs.

Albrecht Dürer's woodcut of the Northern Hemisphere constellations from 1515.

This is because pre-existing stories were projected onto the stars. Tales of heroic princes, damsels in distress, vain kings and magical dragons were told using the night sky as one giant picture book. In the days before the printed word, our stories formed part of a rich oral storytelling tradition. The stars were a way to remember them all. But, more than that, they were a way to pass vital information down the generations.

Our ancient ancestors noticed that, just like the weather,

some constellations come and go with the seasons. The famous constellation of Orion dominates the northern hemisphere sky in the winter, only to go into hiding when the weather takes a turn for the better. By keeping track of these astronomical cues, our predecessors knew the best time to plant and to harvest. Astronomical knowledge was, in effect, a giant agricultural textbook passed from parent to child by means of telling stories about the stars. The constellations made that information all the easier to remember.

Today, professional astronomers officially recognize eighty-eight constellations spanning both hemispheres. The northern hemisphere constellations are largely a legacy of the myths and legends we inherited from the Ancient Greeks and the Romans. Examples include the famous winged horse Pegasus and his rider Perseus. The southern hemisphere constellations were mostly drawn by the first European explorers to chart its unexplored waters. Consequently they are a touch more practical and a little less fanciful. Microscopes, telescopes, nautical equipment, ships, fish and seabirds abound.

Each civilization had its own constellations, from the Australian Aborigines and the Chinese, to the Alaskan Inuit and the Incas. But the eruption of the scientific revolution in Europe led to the adoption of the Greco-Roman constellations as the official global standard. They have changed and been chiselled many times over the centuries, but in 1922 the International Astronomical Union (IAU) officially fixed them in perpetuity.

They remain a useful way of breaking up the night sky, rather than a real feature of the universe. If you were born on a planet orbiting a star in the night sky, and not the sun, you'd still largely see the same stars, but from an entirely different angle. With them appearing in different positions relative to each other, your ancestors would almost certainly have drawn entirely different shapes between them.

The zodiac and the ecliptic

The stars are still there during the day, we just cannot see them because the sun outshines their meagre light. It's like trying to see a candle in the floodlights of an 80,000-seater stadium. Nevertheless, it is possible to talk about the sun residing in a particular constellation, even if we cannot see its individual stars at the time.

Each day the sun appears to move just less than one degree across the sky compared to the background stars. In one year it completes a 360-degree circuit. The path the sun traces out across the sky is known as the *ecliptic*. This did not escape our ancestors' attention. As far back as the first millennium BCE, the Babylonians divided up the ecliptic into twelve constellations – one for each lunar cycle of a normal year. Even if you don't know much about astronomy, it is still likely you'll have heard of the modern version of these constellations: Aries, Taurus, Gemini, Leo, Cancer, Virgo, Libra, Scorpius, Sagittarius, Capricornus, Aquarius and Pisces. These are the twelve constellations of the 'zodiac', which means 'circle of little animals'.

In the past, the night sky was intimately linked with mysticism and superstition. Events in the heavens were often seen as influencing proceedings on the ground. This is the origin of astrology – the idea that the motions and positions of celestial objects affect human affairs. In particular, that the constellation in which the sun resides on the day of your birth has some bearing on the rest of your life. Yet our modern *astronomical* understanding allows us to say that there is zero evidence that this is the case. The stars are just big hot balls of gas very far away. Their positions on the day you were born have as much chance of influencing your life or personality as the position of a vase on your maternity ward shelf or whether your dad's car was parked facing north in the hospital car park.

A sixteenth-century woodcut showing the twelve zodiac
constellations that together trace out the annual path
of the sun across the sky.

However, the zodiac and the ecliptic did play a major role in our move away from this superstitious picture to a more scientific one. As we'll see next, observing the movement of objects close to the ecliptic was crucial in revolutionizing our understanding of our place in the universe and throwing out old, unfounded ideas.

Wandering stars

To the ancients there were three types of star. Those which neatly stayed in their constellations were known as *fixed* stars. Occasionally a *shooting* star appeared, flashing across the sky in a brief moment of brilliance (see page 77). Then there were the *wandering* stars. Just five stars formed this small group of rebels defying the normal rules. They were seen to move close to the ecliptic, entering one zodiac constellation before departing for another. The Greek for wandering star is *asteres planetai*, which is where we get our modern name for these wanderers: planets.

In Europe, these misfits were named Mercury, Venus, Mars, Jupiter and Saturn after members of the pantheon of Roman gods. Along with the moon and the sun, they make up the seven objects that appear to buck the trend and move through constellations. Our ancient ancestors gave their names to the seven days of the week (see table on page 18). The fact that distant civilizations independently came up with a seven-day week suggests that it is because they could all see seven objects moving close to the ecliptic. After all, other time periods are derived from the sky.

The planets Uranus and Neptune also move along the ecliptic, but were unknown to the ancients because they are too far from the sun and therefore too faint to be seen without a telescope. It is interesting to think that if humans had evolved bigger eyes, and therefore the ability to see Uranus and Neptune unaided, we might live in a world with a nine-day week.

English	French	Spanish	Celestial Object
Monday	lundi	lunes	moon
Tuesday*	mardi	martes	Mars
Wednesday*	mercredi	miércoles	Mercury
Thursday*	jeudi	jueves	Jupiter
Friday*	vendredi	viernes	Venus
Saturday	samedi	sábado	Saturn
Sunday	dimanche	domingo	sun

* The English words for these days are derived from Norse/Anglo-Saxon gods, so they don't match the Roman names for the planets.

If you were to observe the planets over months and years you'd notice they appear to do something strange. First they move in one direction along the ecliptic, only to stop, change direction and return from where they came. This is known as *retrograde motion*. Anyone professing to understand the inner workings of the sky had to be able to explain this unusual behaviour.

Ptolemy and the geocentric model

Past civilizations, most notably the Ancient Greeks, pieced together all they knew about the sky to create a model of the universe. They knew the Earth was spherical and that the sun and the stars appeared to rotate around the sky once per day. Their everyday experience told them that the Earth wasn't moving – it certainly didn't feel like it was. So they naturally concluded that we live on a stationary Earth around which the sun, moon, planets and stars revolve. This idea of a central Earth is known as a *geocentric* model.

And it made a lot of sense. Not only did it match what they observed in the sky, but it chimed nicely with religious ideas about the Earth being the centre of creation. Most models at the time had the Earth surrounded by a series of wheels on which sat the sun, moon, planets and stars. As the moon moves the fastest across the sky, it naturally sat in the first wheel. Then came Mercury, Venus, the sun, Mars, Jupiter and Saturn. Beyond Saturn was the wheel belonging to the fixed stars in their constellations.

Early astronomers believed in a geocentric universe in which the sun orbits Earth, shown here in an illustration from 1687.

But there was one major problem with this model – it couldn't easily explain the retrograde motion of the planets. Why would some wheels suddenly stop turning and then begin to turn the other way? The Greek mathematician Claudius Ptolemy came up with a solution that came to be known as the Ptolemaic model. He said that the planets move on a small circle called an epicycle, and that circle moves around the Earth on a larger circle or wheel called a deferent (see picture on page 21). When the planet's motion along the epicycle is in the same direction as the deferent's, we see the planet move one way along the ecliptic. But the planet appears to change direction when it moves around the epicycle against the

CLAUDIUS PTOLEMY (c. 100 – c. 170 CE)

For a man with such an influence over more than a thousand years of astronomical thinking, remarkably little is known about Ptolemy. Only his work persists. He lived in Alexandria, then part of the Roman Empire and now part of Egypt.

In his book *Planetary Hypotheses*, he set out his system of epicycles and also attempted to calculate the size of the universe. He thought the distance to the sun was 605 times the Earth's diameter (it's closer to 12,000). He thought the distance to the stars was 10,000 times the Earth's diameter (it is more than 3 billion times). His other famous astronomical work, *Almagest*, lists forty-eight constellations (as opposed to the modern eighty-eight) – many of them are still in use today.

He was an avid astrologer, although some sources credit him with realizing that someone's life circumstances also have an effect on their behaviour and personality. He wrote works on music, optics and geography, too. As with Eratosthenes, there is a crater on the moon named after him.

direction of the deferent. It was a clever solution, and one that was so accurate in explaining the motions of the heavens that it remained unchallenged for more than a thousand years.

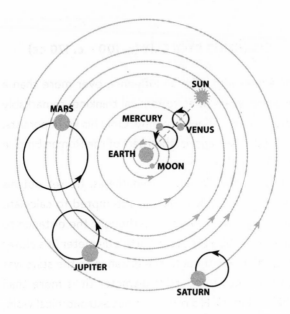

Ptolemy invented deferents and epicycles to
explain the retrograde motion of planets.

Copernicus and heliocentrism

By the sixteenth century, the Ptolemaic model had become so engrained in Western culture that to question it could literally put your life in danger. Christianity had swept across Europe since the days of Ancient Greece and one of its core teachings was that God created the universe in seven days. So it seemed natural that the Earth was the centre of creation – why go to all that trouble and not place your work at the heart of the action? To argue that it wasn't was an act of heresy. Muslim scholars in the Middle East weren't tied so tightly to such dogma and were starting to find cracks in Ptolemy's geocentric idea as early as 1050 CE.

Back in sixteenth-century Europe, a Polish mathematician called Nicolaus Copernicus realized that you don't need epicycles and deferents to explain the retrograde motion of the planets. All you had to do was place the sun in the centre and have the Earth as just one of the planets orbiting it. This is a *heliocentric* model of the universe.

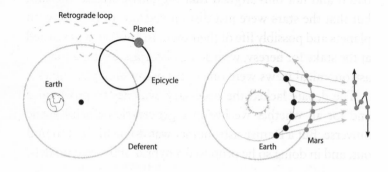

How retrograde motion is explained from Ptolemaic and Copernican viewpoints.

Mars's apparent retrograde motion would then simply be a consequence of us 'lapping' the planet on our journey around the sun. Our motion towards Mars would see it appear to move one way, but once we've hurtled past the planet it would seem to move away from us as we flee. In the first decades of the sixteenth century, Copernicus began to write about his ideas and gave secret copies to trustworthy friends. By 1532 he was sure he was correct, but resisted making his work public for fear of recrimination. It is said – although it is the subject of much debate – that Copernicus only saw a copy of his finished book on his deathbed. According to the story, safe in the knowledge that his ideas would finally be published, he died peacefully in 1543. That work – *De revolutionibus orbium coelestium* (*On the Revolutions of the Heavenly Spheres*) – is arguably one of the most important books ever written.

It caused a theological crisis. By the late sixteenth century, Italian friar Giordano Bruno had picked up the intellectual baton and not only argued that the Earth orbited the sun, but that the stars were just distant versions of the sun with planets and possibly life of their own. In 1600 he was burned at the stake for heresy, with some historians arguing that his astronomical views were one of his many 'thought crimes'.

The debate lacked the necessary evidence to prove once and for all whether we live in a geocentric or heliocentric universe. But a Danish astronomer was doing his best to find out, and in doing so he proposed a hybrid of the two models.

Tycho Brahe

Danish astronomer Tycho Brahe was the dictionary definition of an eccentric. For much of his adult life he sported a brass nose – aged twenty he'd lost the tip of his nose to the point of a sword during a duel over mathematics. Some historians even argue that William Shakespeare based his character Hamlet on Brahe – the characters Rosencrantz and Guildenstern certainly share their names with Brahe's cousins. It is even possible that the whole of *Hamlet* is an elaborate allegory of the battle between the geocentric and heliocentric models of the universe, with the character Claudius named after Claudius Ptolemy.

What we do know is that Brahe's real passion was astronomy and he was very good at it. He made more accurate measurements of the heavens than any astronomer before him. The Danish king gifted him the small island of Hven (now part of Sweden) along with funding to build a giant astronomical observatory there. Brahe named it Uraniborg, meaning the Castle of Urania – the daughter of Zeus and muse of astronomy.

The social calendar at Uraniborg was almost as noteworthy as the astronomical observations made there. Brahe employed a dwarf jester called Jepp, who would often hide under tables and leap out to surprise guests. He also kept a tame moose in the grounds, which came to an unfortunate end when it sipped from an open vat of beer and, intoxicated, fell down the stairs. Brahe would meet an equally unfortunate demise. Attending a lavish banquet in Prague in 1601, he refused to leave the table to visit the toilet despite consuming a large quantity of alcohol. He died eleven days later, killed by uraemia – the condition of having too much urea in the blood. His bladder had burst.

Before his untimely end at the age of fifty-four, Brahe carefully recorded the motion of the stars and the planets from Uraniborg using sextants and quadrants – mechanical devices for measuring angles between objects in the sky. Many of his observations were accurate to 1/60th of a degree. This led him to a compromise between geocentrism and heliocentrism. He was unable to bring himself to believe that something as bulky as the Earth moved, and so his Tychonic model had the sun and the moon orbiting the Earth and the planets orbiting the sun. Like Ptolemy's epicycles, this accounted for the retrograde motion of the planets.

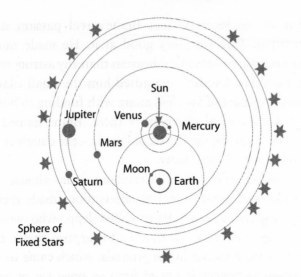

Brahe invented a hybrid model in which Earth is
still central, but some planets orbit the sun.

At least on paper. But there still wasn't enough evidence to
categorically decide which of the three models – Ptolemaic,
Copernican or Tychonic – accurately described the real
universe in which we live. Then an accidental discovery by
a Dutch spectacle maker changed astronomy for ever.

The invention of the telescope

U p until this point, all astronomical observations were
made with the naked eye, sextants and quadrants.
Then Dutchman Hans Lippershey built the first telescope
in 1608, applying for a patent for a device *for seeing things
far away as if they were nearby*. It is unclear whether he
really was the first person to fashion such an instrument, but

history often awards him the credit. Many breakthroughs in the history of science, such as Archimedes' Eureka! moment or Isaac Newton's falling apple, come with stories of the moment of insight – probably apocryphal ones. The invention of the telescope is no different.

It is said that Lippershey's light-bulb moment came when he saw two children playing with a box of old lenses in his workshop. Looking at a distant weather vane through two lenses at once caused it to suddenly appear much bigger. Lippershey used this effect to build a device that could magnify objects by three times. A few years later, Greek scientist Giovanni Demisiani dubbed this new invention a *tele-scope* from the Greek for 'far' and 'to look or see'.

But it was an Italian mathematician who used this new invention to its true potential, and in doing so finally ruled out a very old idea.

Galileo and his telescopic observations

In 1608 Italian scientist Galileo Galilei was working in Padua teaching mathematics at the local university. During a trip to Venice he encountered a copy of a newly invented Dutch device spreading like wildfire across Europe. He worked to improve the design and quickly came up with a telescope with a magnification of eight times (compared to Lippershey's original three). Before long he'd built a device capable of magnifying by more than thirty times.

To Galileo it quickly became clear that we don't live in a completely geocentric universe. Ptolemy was wrong. On 7 January 1609 he directed his telescope towards Jupiter, seeing three small objects around the planet. Within a week he'd noticed a fourth. The four largest moons of Jupiter, they're now nicknamed the Galilean moons in his honour (see page 101). Here were four objects clearly not orbiting either the Earth or the sun.

The real clincher came in September 1610 when Galileo saw that Venus had phases just like the moon. Sometimes it appeared 'full', at other times as a crescent. Venus's size changed, too, as if getting closer to us and then further away. It is not possible for us to observe Venus as having phases if both Venus and the sun orbit the Earth as Ptolemy had suggested. A Ptolemaic system doesn't allow Venus to sit between Earth and the sun – an alignment that has to happen in order for us to see the phases. Under the Tychonic and Copernican systems, when Venus lies between us and the sun, we hardly see it illuminated at all because most sunlight falls on the opposite side of the planet. The side facing us is completely lit up when it is the furthest from us.

At last, here was evidence to rule out Ptolemy's ancient geocentric model. But to come down on the side of heliocentrism could still land you in hot water. When Galileo argued in favour of Copernicus he incurred the wrath of the clergy. They advocated Tycho's system because it matched both the phases of Venus and the religious need for a central Earth. In 1616, an Inquisition declared the idea of heliocentrism to be in direct contradiction to Holy Scripture. In 1633, Galileo was brought to trial and found guilty of heresy. His punishment was a life under house arrest. He passed his days writing important books on less controversial areas of science until he died aged seventy-seven in 1642. The Church eventually pardoned Galileo – but only in 1992!

Galileo also drew pictures of the moon's mountains and used the length of shadows to estimate their height. His findings revealed a world with taller peaks than anyone had expected. The first to see Saturn's rings, he described them as 'ears' sticking out on either side of the planet. He even observed spots on the surface of the sun and revealed that our Milky Way is not just a cloud of gas but densely packed with stars.

Johannes Kepler and his planetary laws

German mathematician Johannes Kepler was one of the Copernican model's earliest and most vociferous advocates, even before Galileo's observations. Made an assistant to Tycho Brahe in 1600, Kepler yearned for the mathematical rules by which the planets orbit the sun. He was allowed to use some of Brahe's observations, but the Dane guarded his data closely. The fact that Brahe died just a year later, and Kepler conveniently inherited all of his work, has led some historians to call foul play. Brahe's body was exhumed in 1901 and traces of mercury were found in his remains. Did he really die of bladder failure? Or did Kepler poison him to get at his unrivalled astronomical catalogue? After all, it's only from Kepler's diary that we know the story of Brahe's death. However, his body was exhumed again in 2010, with tests revealing mercury levels insufficient to have caused his demise.

In the decade after Brahe's passing, Kepler used his observations to form his famous three laws of planetary motion:

Kepler's First Law: *Planets orbit the sun in ellipses, with the sun at one of the foci.*

He could see that the planets do not go round the sun in perfect circles as the ancients and even Copernicus had imagined. Instead they trace out an oval shape called an ellipse. An ellipse has two 'foci' (singular 'focus') – mathematically important points inside the curve. The sun sits at one of these points.

Kepler's Second Law: *The line between the sun and a planet sweeps out equal areas in equal times.*

A consequence of planets having elliptical orbits is that they are closer to the sun at some points than at others. However, Kepler noticed that a line between the sun and a planet takes the same time to move through the same overall area (see below). Put simply, a planet speeds up when closest to the sun and slows down when furthest away.

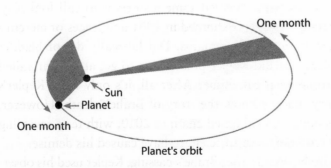

According to Kepler, planets orbit the sun in ellipses and speed up when closer to the sun.

Kepler's Third Law: *The square of a planet's orbital period is proportional to the cube of its distance from the sun.*

In essence, the further a planet is from the sun the longer it takes to go around. This is common sense really – Mercury goes round the sun the fastest because it has the smallest ellipse to navigate. Saturn takes much longer because it has much further to travel. Kepler's big insight was the exact mathematical relationship between these two things. Using Brahe's accurate observations, he noticed that the time of a planet's orbit 'squared' (multiplied by itself) was linked to its distance from the sun 'cubed' (multiplied by itself twice).

These are *empirical* laws – those based on direct observations, not on some underlying theoretical explanation of why the planets orbit the sun. That deeper understanding would come in 1666 when, after being forced to leave Cambridge due to the plague, an English mathematician was reportedly sitting in his mother's garden and an apple fell on his head.

Isaac Newton and gravity

It seems the story of Newton and the apple has some grain of truth to it, but it never hit him on the head. At least not according to an influential biography entitled *Memoirs of Sir Isaac Newton's Life* (1752). The author of that biography – William Stukeley – was drinking tea with Newton in a garden after dinner when the famous scientist told him that he'd hit upon his theory of gravity after seeing an apple fall to the ground.

Newton's key insight was that every object experiences a force of attraction towards every other object in the universe. The apple is attracted to the Earth and so it falls. The only reason it stops falling is that it hits the ground. Newton realized that if you could propel the apple high enough, and with sufficient speed, it would fall around the Earth without the planet's surface getting in the way. It would orbit the Earth. His big jump in reasoning was that the moon orbits the Earth for the same reason the apple drops – it is in freefall with nothing in the way. All because of a gravitational attraction between the two objects.

Newton published his ideas about gravity in a book called *Philosophiæ Naturalis Principia Mathematica* (often abbreviated to simply *Principia*) in 1687. It also contained many other hugely important insights, including his famous three laws of motion. Newton states in the book that the gravitational attraction between two objects is inversely proportional to the square of the distance between them. This means that if you double the separation of two objects the strength of their gravitational attraction drops to a quarter. Treble that distance and it drops to a ninth. What made his ideas so powerful is that he used both his Universal Law of Gravitation and his laws of motion to prove Kepler's Laws of Planetary Motion (see page 28). He was effectively saying, 'I know why the planets orbit the sun and I can prove it because my ideas give the same results as Kepler.'

Take Kepler's Second Law – that a line between the sun and a planet sweeps out equal areas in equal times. Said another way, planets speed up when closest to the sun and slow down when furthest away. Newton provided an explanation for this behaviour. The gravitational attraction between two objects gets stronger the closer they are together and weaker the further they are apart. When a planet is close to the sun it feels a stronger pull and is sped up; as it moves away the strength of that force decreases and the planet slows down.

But Newton's magnum opus nearly didn't make it to print. The Royal Society had blown their publishing budget on a flop of a book called *The History of Fishes*. So astronomer Edmund Halley stepped in and paid for its publication himself. In doing so he ensured that one of the most important books of all time, scientific or otherwise, saw the light of day.

Isaac Newton and light

Around the same time that the falling apple sparked Newton's imagination, he was also busy playing with prisms and light. Experimenting with these glass blocks was nothing new, and it had long been known that they could produce a range of colours from white light. But the prevailing view was that the prisms themselves coloured the light as it passed through them. Light itself was pure white.

Newton disproved this idea through a clever but simple experiment. One sunny day in 1666 he closed a blind over a window and poked a small hole in it so that a single beam of sunlight entered the room. He placed a prism in the path of the light so that a rainbow of colours appeared as expected. Now for the clever bit. He placed a second, inverted prism in the path of those colours. Sure enough, the second prism recombined the colours back into white light. Prisms can't be adding colour after all. White light must really be a mix of different colours that prisms can separate (or recombine). Newton published his results in 1672.

THE REFLECTING TELESCOPE

Newton designed a new type of telescope in 1668. Early telescopes were *refractors* – they used lenses to bend or *refract* light. But Newton's reflecting (mirrored) telescope avoided one of the biggest problems with refractors – *chromatic aberration*. Lenses bend each colour of light slightly differently just as in a prism, meaning they're not all brought to the same focus.

In a Newtonian telescope, light enters at the top and strikes a curved mirror at the bottom. Light is reflected back up the tube, hits a flat, secondary mirror and is bounced out to the side where an eyepiece reveals the focused image.

Today, the biggest telescopes are reflectors because there's a limit to the size of a refractor. Light must pass through the lens, which means supporting it on either side. Build it too big and it sags under its own weight and no longer adequately focuses light. A mirror, however, can be supported from behind. The world's largest refractor has a one-metre lens – the largest reflector has a diameter of just over ten metres.

This basic understanding of the properties of light is the bedrock of many areas of modern astronomy. As we'll see in later sections, astronomers rely on it time and time again.

Römer and the speed of light

The late seventeenth century was a truly revolutionary period in our understanding of light. Not only was Isaac Newton making valuable discoveries about the origin of colour, Danish astronomer Ole Römer was working out how fast light travels.

In the 1670s, the Royal Observatory in Paris despatched a team of astronomers to Tycho Brahe's old Uraniborg observatory on the island of Hven to make careful measurements of the four Galilean moons of Jupiter. Specifically to note when they disappeared from view as they were eclipsed by the planet. Römer was the local assistant to the French astronomer Jean Picard and was subsequently offered a job in Paris on the back of his work at Uraniborg.

Observations of the moons threw up a puzzling mystery: eclipses often occurred either earlier or later than calculations based on Newtonian gravity predicted. By 1676, Römer had hit upon the explanation, building on the work of the Observatory's director Giovanni Cassini. They correctly suggested that it takes light time to travel across space. Previously it was thought that the speed of light was infinite – that it got from A to B instantaneously. But the eclipses of Jupiter's moons appeared to happen ahead of schedule if Earth and Jupiter were close together, and were apparently delayed when the two planets were far apart. Römer calculated that it takes light eleven minutes to cross a distance equal to Earth's distance from the sun. That's a speed of 220,000,000 (220 million) metres per second.

Today we know that the speed of light is 299,792,458 metres per second, so Römer and Cassini weren't too far out. The important thing is not the number they came up with, but the fact they could conclusively show the speed of light to be finite – it takes light time to get places. In our everyday lives we never notice because it is so swift. Only over astronomical distances does it become noticeable. We'll return to this idea many times.

One well-known way to speak of cosmic distances is to talk in light years – the *distance* light travels in one year. Travelling at 299,792,458 metres per second, light travels 9.46 trillion kilometres in a year. The nearest star to the Earth after the sun sits about 40 trillion kilometres away – or 4.2 light years. For closer objects you can use light hours, light minutes or even light seconds. Pluto, for example, is 5.3 light hours from the Earth. The sun is 8.3 light minutes away, and the moon just 1.3 light seconds.

Halley and his comet

In the 1670s, both the French and English kings set up Royal Observatories with the aim of using the stars to aid navigation at sea. In England, the Director of the Royal

Observatory in Greenwich was given the title Astronomer Royal. When John Flamsteed, the first Astronomer Royal, died in 1719, the post passed to his assistant Edmund Halley – the man who had privately funded the publication of Newton's *Principia* (see page 32).

Part of the reason Halley paid to have *Principia* published was that he had personally seen the power of Newton's work. In 1684, three years prior to the book's publication, Halley visited Newton and the two men discussed gravity and how it relates to comets – icy lumps of debris tumbling in orbit around the sun (although that wasn't widely known at the time). In 1680, a spectacular comet called Kirch had blazed across the sky. Newton used Flamsteed's observations of the comet to show that it, too, obeyed Kepler's Laws – its orbit was elliptical and it sped up on approach to the sun – so, like the planets, it must be affected by the sun's gravity.

By 1705, Halley had built on Newton's work and produced his own book on comets called *Synopsis of the Astronomy of Comets*. Now knowing that comets orbited the sun, he suggested that three comets, appearing in 1682, 1607 and 1531, were actually the same comet making return journeys by the Earth on successive orbits. He predicted that it should return again in 1758. Halley died in 1742 and so did not live to see the comet's return in the very year he'd forecast. The object has since been known as Halley's Comet in his honour.

Armed with this new knowledge, astronomers and historians looked back over history and found records of the same comet spanning generations and continents. Comets observed in fifth-century BCE Greece and third-century BCE China bear all the hallmarks of Halley.

Famously, it even appears in the Bayeux Tapestry. Halley's Comet last visited the inner solar system in 1986 and is due for a return in 2061.

Bradley and the aberration of light

Despite the successes of Galileo, Kepler, Newton and Halley, the debate between the Tychonic and Copernican models raged on. There still wasn't undeniable proof that the Earth was in fact moving around the sun.

Both Picard in Paris and Flamsteed in Greenwich noticed that the Pole Star – the star which seems to stay in the same spot no matter what the season – actually moves back and forth slightly over the course of a year. It would take James Bradley, Halley's successor as Astronomer Royal, to provide a concrete explanation and with it bury all geocentric models.

Imagine starlight falling onto the Earth as rain. When you walk into vertically falling rain it appears to strike your umbrella at an angle from the side. The rain isn't

When you move into falling rain it appears to strike your umbrella from an angle.

really falling at an angle – it is your motion into the rain that creates the effect. Similarly, Earth travels one way into the 'rain' of starlight during one half of its orbit, and moves into it from the opposite direction during the other half. It is this effect – known as *aberration* – that makes stars appear to move slightly in the night sky over the course of a year. In a Tychonic system, with a stationary Earth, there would be no such effect. Finally, in 1729, Bradley gave us cast-iron proof that we live in a helio-centric, Copernican solar system. However, the Catholic Church continued to ban all books on heliocentrism until 1758.

The Transit of Venus

With Earth established as just another planet, astronomers' attention turned to exactly how far we are from the sun. The only way to measure this distance in the eighteenth century was by observing a rare astronomical event called the Transit of Venus, which occurs when Venus passes directly in front of the sun from our perspective, a bit like a mini solar eclipse.

If you were to observe the transit from two different locations on Earth – the further apart the better – you'd see the event begin and end at slightly different times because you'd be viewing the sun from slightly different angles. Halley realized that you can use this time difference to work out the distance between Earth and Venus. Kepler's Third Law can then be used to scale this up to the Earth–sun distance.

However, with the planet seeming small due to its large distance from us, these events are not easily visible without a telescope. They happen in pairs, with eight years between each transit, but then you have to wait over a century for the next pair.

Johannes Kepler used his laws of planetary motion to predict that a transit would occur in 1631, the first such prediction. He was right; however, it happened during European night and nobody was able to see it. English astronomer Jeremiah Horrocks correctly predicted another transit in 1639 and in observing it from his home near Preston became the first person ever to see one. Edmund Halley hit upon the method for using these events to calculate the distance to the sun in 1691, but astronomers had to wait until the next two transits in 1761 and 1769 to make a concerted effort to use it.

Such was the importance of this measurement – and the scarcity of opportunity to make it – that eighteenth-century astronomers went to great lengths to ensure they didn't miss their twice-a-century chance. European observatories despatched teams of astronomers right around the globe to make observations of the 1761 and 1769 transits and to multiple locations to ensure the weather didn't hamper their work. If one team was clouded out, another might have clear skies.

The Royal Society commissioned HMS *Endeavour* under the command of Captain James Cook to sail to Tahiti to observe the 1769 eclipse. Cook also carried with him sealed orders from the British government for what to do after the transit – they told him to search the Pacific for a rumoured undiscovered continent. On 29 April

1770 he famously landed at Botany Bay (in modern-day Sydney) and the site became the first European settlement in mainland Australia.

The measurements of the Transit of Venus from Tahiti were used to infer an Earth–sun distance of 93,726,900 miles (150,838,824 kilometres). Today's value is 149,600,000 kilometres, so the eighteenth-century astronomers were remarkably close given their limited technology.

Weighing the world

Astronomers also wanted to know how heavy the planets are. In the eighteenth century, even the mass of the Earth was a mystery. Edmund Halley, for all his comet success, thought the Earth was hollow. In 1774, one of his successors as Astronomer Royal – Nevil Maskelyne – showed that it wasn't.

Since the days of Newton's *Principia* we've known that every object experiences a gravitational attraction towards every other object in the universe. The closer the objects, the stronger the attraction. Newton himself had considered the possibility of using this to weigh the Earth. He envisioned a pendulum held close to a large mountain. The bob on the end of the pendulum experiences three forces: a gravitational pull towards the mountain, a gravitational pull towards the Earth and the tension in the string holding it up. The result is that the bob rests at a slight angle to the vertical in the direction of the mountain. Here the joint pull of the mountain and the Earth match the strength of the tension in the string. If you measure the mass of the mountain, and the angle by which the bob is deflected, you can use Newton's equations to calculate the mass of the Earth.

Newton ruled out the experiment on practical grounds, thinking it would be too difficult to measure the deflection of the bob. But Maskelyne took on the task. He selected the 1,083-metre peak of Schiehallion in Scotland due to its symmetrical, conical shape. It is relatively easy to calculate the volume of a cone, and if you know the density of rock that the mountain is made of then you can work out its mass. Maskelyne set up observing points on either side of the mountain and, despite setbacks due to terrible weather, eventually managed to measure the deflection angle of the pendulum using the stars as a reference point. Surveyor Charles Hutton then began to work out the volume of the mountain. To help his efforts he split it into sections and in doing so invented contour lines.

Maskelyne's team calculated an average density of the Earth of 4.5 grams for every cubic centimetre (the modern value is 5.5 g/cm^3). Given that the average density of the Schiehallion mountain was just 2.5 g/cm^3, there must be significantly heavier material than the mountain under the surface of the Earth – our planet cannot be hollow. Up until this point the densities of the sun, moon and planets had only been known as multiples of Earth's. Once the Earth's average density was known, astronomers could then say something about the densities and masses of all the other large objects in the solar system too. One mountain in Scotland effectively set the scales for the sun's entire family of orbiting worlds.

Object	Mass	Density
Earth	5.97×10^{24} kg	5.5 g/cm^3
	Compared to Earth	
Sun	333,000	
Moon	0.01	0.61
Mercury	0.06	0.98
Venus	0.82	0.95
Mars	0.11	0.71
Jupiter	317.8	0.24
Saturn	95.2	0.13
Uranus	14.5	0.23
Neptune	17.1	0.30

Herschel and Uranus

On 13 March 1781 William Herschel doubled the size of the known solar system overnight. From his home in Bath, England, he had discovered an entirely new planet twice as far from the sun as Saturn. As all the other planets had been known since antiquity, it was the first time a planet had ever truly been 'discovered'. It turns out that many astronomers – including several Astronomers Royal at Greenwich – had seen it before, but because it moves so slowly along the ecliptic it had always been mistaken for a fixed star. At first Herschel thought it was a comet, but he gradually realized its true nature.

However, it took nearly a century to universally agree on a name for this new find. As discoverer, Herschel had initial naming rights and chose *Georgium Sidus* (or George's Star) in recognition of King George III, who'd employed Herschel as an astronomer. You can imagine that this name wasn't quite as popular in other countries. In 1782, Uranus – the Greek god of the sky – had been suggested as a neat alternative as Uranus was the father of Cronos (Saturn) who in turn was the father of Zeus (Jupiter). But it took until 1850 before this name was officially adopted. The name does make the planet stand out. All other planets (besides Earth) are named after Roman gods – Uranus is the only one with a Greek name.

Herschel and infrared light

In 1800, Herschel made a discovery arguably even more important than finding a new planet: he found an entirely new form of light.

Like Newton over a century earlier, Herschel was experimenting with prisms. He had come to suspect a link between colour and temperature. So he passed sunlight through a prism and placed thermometers at different positions in the colour spectrum it produced. He found the highest temperatures at the red end of the spectrum. But then he did something remarkable: he moved the thermometer past the red band of light to a place where there appeared to be no light at all. His thermometer recorded an even higher temperature in this region than anywhere in the colour spectrum.

Herschel said there must be some invisible 'calorific rays' beyond the red end of the light spectrum. His subsequent experiments with these rays showed that they behaved in exactly the same way as ordinary rays of light. We now know his calorific rays as infrared radiation. It is the invisible light emitted by hot objects – which is why modern infrared cameras are used to pick up heat signatures in police chases, battlefields and disaster zones.

Herschel's discovery was the first indication of light beyond what our eyes can see. Just as there are sounds with frequencies too low or high for the human ear to hear, so there are frequencies of light too low or high for our eyes to see. Modern physicists refer to the full range of light frequencies as the *electromagnetic spectrum*. It ranges from radio waves and microwaves at the low-frequency end, up through infrared and visible light to ultraviolet, X-rays and gamma rays. An astronomer would call all these things light.

The first telescopes were all sensitive to the same light that our eyes can see – visible light. But modern astronomers have a suite of telescopes at their disposal capable of seeing in all frequencies of light, from radio waves to gamma rays. If we restricted ourselves only to visible light then we'd be missing out on a lot of the information arriving at the Earth from space.

When the European Space Agency launched the largest-ever infrared space telescope in 2009 they named it *Herschel* in recognition of his great insight.

The discovery of Neptune

I f Uranus was stumbled upon by accident, then Neptune was discovered by design. Astronomers scrutinized Uranus's orbit carefully in the decades after its discovery and spotted some irregularities. The planet didn't always appear to be in the spot where Kepler's and Newton's equations said it should be.

However, it was quickly realized that those laws were not at fault. What astronomers were witnessing was another, more distant planet affecting Uranus's orbit. As Uranus approaches this unseen planet it is pulled towards it and speeds up. Once it has moved past, the planet tries to pull it back the other way and so Uranus slows down slightly.

French mathematician Urbain Le Verrier used Kepler's and Newton's equations to calculate where this meddling planet must be situated. Le Verrier then sent his calculation to German astronomer Johann Galle in Berlin, who pointed his telescope at those co-ordinates and found Neptune waiting for him (it was within one degree of where Le Verrier said it would be). In hindsight, like Uranus it had been spotted several times before (including by Galileo) but its slow speed had made it indistinguishable from a fixed star.

Einstein and special relativity

I t's the most famous equation in all of science. In 1905, $E=mc^2$ appeared when Albert Einstein published his Special Theory of Relativity. It says that energy (E) is equivalent to mass (m). To work out how much energy is locked up in an object you multiply its mass by the speed of light (c) squared.

That year saw a flurry of work from Einstein – he published two other landmark papers. One would later earn him the 1921 Nobel Prize in Physics for discovering that light is made of particles called photons. His insights were especially remarkable given that he was an academic outsider at the time, working instead as a patent clerk in Bern, Switzerland.

Special relativity takes Ole Römer's work on light (see page 34) a step further. Einstein not only said light has a finite speed, but that it's also the cosmic speed limit. Nothing can travel through space faster than light. This idea naturally flows from $E=mc^2$. The faster you move, the more energy you have. But the equation tells us that a gain in energy also means a gain in mass, so the faster you move the heavier you get. Heavier objects are harder to move and so require more energy to boost their speed. If they do travel faster, they get heavier again. Eventually a fast-moving object becomes so heavy that it would require an infinite amount of energy to make it go any faster. This cut-off point is the speed of light.

Einstein and general relativity

Not content with gifting the world his Special Theory of Relativity, Einstein published his General Theory of Relativity in 1915. In doing so he revolutionized our ideas about gravity.

Newton thought of gravity as a pull exerted across

empty space by massive objects. To him, that's why the Earth orbits the sun. Einstein suggested instead that Earth orbits because the sun is changing the shape of space around it. Einstein bundled up the three dimensions of space and the one of time into a four-dimensional fabric he called *space-time* and said massive objects warp it.

The classic way to picture this is to imagine space-time as a bedsheet held tight at the corners. Place a bowling ball in the centre to represent the sun and the sheet will sag, creating a depression – or well – in the middle. Take a tennis ball to represent the Earth and roll it around the rim of the well and it will orbit the bowling ball.

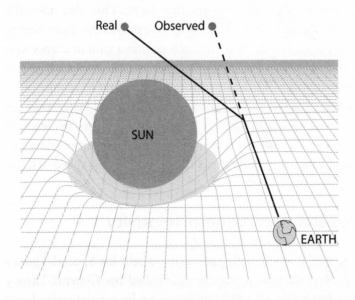

Einstein said that massive objects warp a four dimensional fabric called space-time and that this can bend light from distant stars.

ALBERT EINSTEIN (1879–1955)

No scientist before or since is as famous as Albert Einstein. His face adorns clothes, posters and mugs around the world. His work is as relevant today as it's ever been and physicists are still finding evidence that he was right over a hundred years since his special and general theories of relativity were published. Rightly or wrongly, his grey haired, mad-professor persona has become the stereotypical image of the genius scientist.

He certainly led a colourful life. In 1903, he married his fellow physics student Mileva Marić, but later he started an affair with his first cousin Elsa. Elsa and Albert later married in 1919 and stayed together until she died in 1936. Albert was said to be heartbroken.

German and Jewish by birth, he stayed in America once Adolf Hitler came to power and became a US citizen in 1940. In 1952 he was offered the post of President of Israel but declined. He died in 1955 of an aneurysm and his brain was removed without permission during the autopsy for further studies linked to intelligence.

Astronomers had long known that Newtonian gravity cannot explain oddities in Mercury's orbit. When Einstein applied his curved space-time idea to Mercury it was a perfect fit. But to be sure we needed another way to test it. The secret was to harness the unique circumstance of a solar eclipse.

Both Einstein and Newton agreed that the sun's gravity bends light from distant stars, but disagreed on how much. So, in 1919, British astronomer Arthur Eddington was sent to the tiny African island of Príncipe to find out. Normally you can't see the stars close to the sun in the daytime sky. Yet during a solar eclipse the moon handily blocks out the sun's glare. Eddington used this opportunity to take photographs of the stars near the sun.

Sure enough, the stars were precisely where Einstein said they'd be, displaced from their normal position exactly as they would be if their light had followed a curved path caused by the sun's warping of the space-time around it (see page 47). General relativity remains our best theory of gravity and has so far passed with flying colours this and every other test thrown at it.

CHAPTER 2

The Sun, the Earth and the Moon

The sun

What is it made of?

How do you work out what something just under 150 million kilometres away is made of? Particularly when it is so hot and bright that you clearly can't get anywhere near it without getting severely singed? As with most things in astronomy, the answer lies in the light we receive.

In Chapter 1 we saw how a prism can be used to split white light into a spectrum of colours (see page 32). In the early 1800s, German physicist Joseph von Fraunhofer found that the colour spectrum of the sun is not continuous – it contains a series of more than 500 black lines now known as Fraunhofer lines. In the 1850s, German scientists Robert Bunsen and Gustav Kirchhoff explained why they are there. The lines are simply missing colours – gaps where different substances in the sun have swallowed that particular frequency of light and prevented that colour from reaching Earth.

These lines, in effect, represent a chemical barcode encoding vital information about what the source of the light is made of. It's the sun's unique fingerprint. By heating different elements in laboratories, Bunsen and Kirchhoff matched these 'absorption lines' to those in the solar spectrum (Bunsen invented the burning device which bears his name for this purpose). The sun was found to be mostly made of hydrogen – the lightest element in the universe.

But in 1868 the sun threw astronomers a curveball. That year French astronomer Pierre Janssen observed a solar eclipse and discovered an absorption line that did not correspond to any known element. The same year, English astronomer Norman Lockyer found an identical line when observing the sun. Lockyer and his chemist colleague Edward Frankland named the new element *helium* after 'helios', the Greek word for the sun. Later discovered on Earth, it was the first element to be found in space first. Thanks to this method of spectral line analysis – known as spectroscopy – today we know the sun is 73 per cent hydrogen, 25 per cent helium with the rest made up of other elements including oxygen, carbon and iron.

What powers it?

The sun burns our skin from almost 150 million kilometres away. What powers such an immense furnace was one of the most pressing issues in late nineteenth-century physics.

Developments in geology and biology – including Charles Darwin's work on evolution by means of natural selection – offered hints of a very old Earth. With the sun even older, understanding its power became even more troublesome. It's one thing finding a process capable of sustaining the sun for millions of years. It's another if the sun is *billions* of years old.

Many Victorian scientific luminaries flat out refused to believe in such a long time frame. Lord Kelvin – a leading expert in heat and energy – saw gravity as the sun's power source. As solar material is crushed towards the sun's core, the pressure and temperature increase. This conversion of gravitational energy into heat energy was Kelvin's explanation. However, he calculated the sun would use up all this energy in around 30 million years. It must be younger than that if it is still shining and so in 1862 he publicly rejected Darwin's calculations of an Earth billions of years old.

But Darwin was right and Kelvin was wrong. The missing piece of the puzzle fell into place in 1905 when Einstein published his famous equation $E=mc^2$ (see page 45). It says energy (E) and mass (m) are effectively the same thing and you can swap one for the other. Multiplying a mass by the speed of light (c) squared tells you the amount of energy available. There's a catch, however: liberating energy from mass requires extremes of temperature and pressure.

In 1920, British astronomer Arthur Eddington first described the real mechanism by which the sun is powered: fusion. Helium can be made by fusing hydrogen together under extremes of pressure and temperature such as those found in the sun's core. Crucially, however, the mass of the resulting helium is slightly lighter than the original hydrogen. This missing mass is the source of the sun's power – it's converted into energy according to Einstein's famous equation. Every second, the sun fuses 620 million tonnes of hydrogen into 616 million tonnes of helium. The missing 4 million tonnes is converted into sunshine.

ARTHUR EDDINGTON (1882–1944)

Eddington was one the most important figures in early twentieth-century astronomy. Born in north-west England to Quaker parents, he was about to apply for conscientious objector status to avoid fighting in the First World War when he was granted immunity from conscription due to the importance of his astronomical work.

When Einstein published his General Theory of Relativity in 1915 – in German, during wartime – Eddington was one of the few astronomers capable of understanding it and worked to spread its key ideas to English-speaking academics. Eddington's test of general relativity through a 1919 eclipse made Einstein a household name. Eddington went on to make important contributions to our understanding of the life cycle of stars, including calculating the Eddington limit – the maximum brightness a star can achieve for its size.

He didn't get everything right, however. In the 1930s, Indian astrophysicist Subrahmanyan Chandrasekhar used general relativity to suggest the existence of black holes – an idea Eddington publicly rubbished. Chandrasekhar never forgot the snub but was eventually vindicated, winning the 1983 Nobel Prize in Physics.

Despite the sun's voracious appetite for hydrogen, it still has 5 billion years' worth of material to fuse. We'll see what happens after it runs out of fuel in Chapter 4.

The exact way hydrogen becomes helium was uncovered in 1939 when German-American nuclear physicist Hans Bethe published a blueprint for the proton-proton (pp) chain in which four protons (hydrogen nuclei) eventually fuse into the nucleus of a helium atom. Although this process happens approximately 90 trillion trillion trillion times a second in the sun's core, it can take millions of years for individual protons to fuse.

The solar neutrino problem

We can't see into the heart of the sun and observe the pp chain in action. However, we can predict how much energy the sun should be emitting if that's what powers it. And the two numbers match.

Yet there was a persistent, nagging problem plaguing astronomers into the twenty-first century: there weren't enough solar neutrinos arriving at Earth. Neutrinos are tiny, almost massless subatomic particles. They're also a by-product of Bethe's pp chain and stream out of the sun, flooding the solar system. But they're incredibly anti-social, largely passing straight through ordinary matter like ghosts. Every second there are more solar neutrinos passing through each square centimetre of your body than there are people on Earth. Yet they do you no harm.

Since the 1960s, physicists have built elaborate experiments to detect just a proverbial handful of these particles as they pass through our planet. They quickly noticed that there weren't enough of them arriving. Only about a third of the neutrinos predicted by the pp chain were being picked up. One proposed explanation was that neutrinos shape-shift – change 'flavour' – into two other types of neutrino on their way to the Earth. So the early neutrino experiments – only sensitive to one type of neutrino – missed out on the other two. That's why they only saw a third of what they expected to see.

Between 1998 and 2006, experiments in America and Japan showed that there are indeed three neutrino flavours and that an individual neutrino can switch – or *oscillate* – between them. Taking neutrino oscillation into account, the number arriving at the Earth is exactly what you'd expect if the pp chain is the power source of the sun.

The epic journey of sunlight

Imagine cutting the sun in half to see its layers. Right in the centre you'd see the core, which takes up approximately the inner quarter. Here the gravitational pressure exerted by the material above raises the temperature and pressure to levels high enough to fuse hydrogen into helium via the pp chain. The temperature is a staggering 15 million degrees Celsius; the pressure is so great that material in the core is more than thirteen times denser than lead.

Light moves out from the core into the radiative zone, which extends to 70 per cent of the sun's width. The temperature gradually drops as you move out from the core, until it reaches about 1.5 million degrees Celsius at the top of the radiative zone. While the density of material slowly decreases too, particles are still jam-packed together close to the core. On average, a particle of light cannot make it more than one centimetre before it bounces into something and is knocked off course.

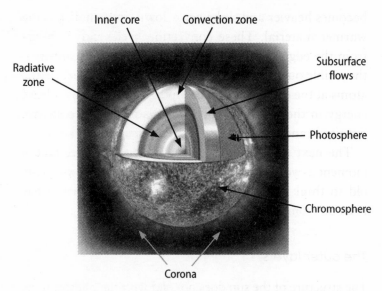

The sun has a multitude of layers from the central core to the outer corona.

If you could follow the path of an individual light particle (photon), you would have to wait between 100,000 and a million years to see it emerge from the crazy, pinball machine environment of the radiative zone. You often hear people say that the light we see from the sun is eight minutes old because that's its journey time from the sun to the Earth. That's the travel time from the *edge* of the sun, but the light isn't created at the edge but in the core. By the time the light reaches our eyes it is already over 100,000 years old.

Passage through the convective zone is much swifter. Typically it takes just three months for the energy to make it out. Once light reaches the convective zone it's absorbed by the gas. This heats the gas, making it lighter and so it rises higher towards the edge of the sun. There it cools, becomes heavier and sinks back down, in turn displacing warmer material. These convection cycles carry energy from the edge of the radiative zone to the photosphere – the visible outer layer of the sun which we can see. As the atoms at the edge of the convective zone cool, they release energy in the form of light, which is now free to stream outwards and illuminate the solar system.

The next time you feel sunlight on your face, take a moment as you bathe in light that is up to a million years old to think about the colossal journey that energy has been on from the core of the sun.

The outer layers

The structure of the sun does not end with the photosphere. There are a few, much more tenuous regions further out – namely the chromosphere and the corona. The chromosphere is home to 500-kilometre-long jets called spicules. There are hundreds of thousands of them on the sun at any one time.

Temperatures keep on dropping as you move out from the core to the photosphere, but then suddenly start to rise as you move away from the photosphere, reaching 8,000 degrees Celsius at the top of the chromosphere. They continue to rise through a narrow, 100-kilometre-wide corridor called the transition zone, climbing to 500,000 degrees Celsius at the base of the corona. Temperatures in the corona reach millions of degrees Celsius. No one really knows why the sun suddenly starts getting hotter again – this *coronal heating problem* is a major topic of modern solar research.

It means solar physicists want to study the corona as much as possible, but its delicate nature is normally outshone by the glare of the layers below. Traditionally, we had to wait for total solar eclipses when the moon conveniently blocks out the rest of the sun. However, many modern solar space telescopes are equipped with coronagraphs – discs that block out the sun to create artificial eclipses so astronomers can routinely study the corona.

These eyes on the sun don't just detect visible light. They are also sensitive to other parts of the electromagnetic spectrum, including ultraviolet (UV) and X-rays. Such observations have revealed coronal holes – dark polar regions that don't emit much radiation. They can persist for months at a time and are the source of the high-speed solar wind (see page 64).

Magnetic fields and differential rotation

The sun is far from the unchanging yellow ball that appears in our sky. It is incredibly dynamic and violent, with a seething surface constantly being shaped and sculpted by intense magnetic activity.

The sun is a giant magnet. You may remember from school that worn-out experiment with a bar magnet and iron filings. The filings align with the invisible magnetic field lines running between the magnet's north and south poles. Both the sun and the Earth have a similar magnetic field running pole to pole. Our magnetic field is reasonably similar to that of the bar magnet because the Earth rotates as one solid planet (see page 74). The sun, however, is a churning, roiling ball of super-heated gas called plasma. Not one solid body, the sun rotates about 20 per cent faster at the equator than at its poles. Astronomers call this *differential rotation*.

The upshot is that the equatorial magnetic field gets dragged on faster, ahead of the field at the poles. This leads to the sun's overall magnetic field becoming a lot more complex as it gets tangled and twisted. Much like coiling a spring, or twisting a rubber band, this process stores up energy in the magnetic field lines. We see the release of this pent-up energy as marks and eruptions from the solar surface.

Sunspots

Sunspots are one of the most obvious features on the sun – dark blemishes often appearing in groups. Galileo first observed them with a telescope way back in the early seventeenth century, but records of sunspots visible to the unaided eye stretch back more than two thousand years. That's possible because some sunspots grow to more than 10 per cent of the sun's overall diameter – or 160,000 kilometres across. That's 12.5 times wider than the Earth. Sunspots typically last between several days or weeks, but the most persistent can remain for months.

Explanations of their origin have varied over the years, from storms in the sun's atmosphere to bruises inflicted by kamikaze comets. Today, we know that they are simply cooler regions in the photosphere. The average temperature of the photosphere is around 5,500 degrees Celsius, whereas the temperature of a sunspot is largely between 3,000 and 4,000 degrees Celsius. In sunspot regions, intense local magnetic fields prevent as much heat as usual from rising upwards from the convective zone below. That's why sunspots often appear in pairs – one for each magnetic polarity.

Since Galileo's time, astronomers have kept detailed records of sunspot numbers. And there's a distinct pattern – the number of sunspots appears to reach a crescendo after roughly eleven years, before dying away and slowly

ANNIE MAUNDER (1868–1947)

Born in Northern Ireland, Maunder (née Russell) studied at Cambridge before becoming one of the 'human computers' at the Royal Observatory Greenwich. She was tasked with taking photographs of the sun and performing calculations. During her time in Greenwich she met fellow astronomer Walter Maunder and they married in 1895. The social expectations of the age meant she was forced to officially give up her job when she married.

However, the pair continued to work together on understanding the sun and sunspots in particular. They examined historical sunspot records and noticed a correlation between low sunspot numbers and periods of lower temperatures on Earth. A period between 1645 and 1715 is known as the Maunder Minimum, or more colloquially as the Little Ice Age.

Maunder – a great communicator of astronomy to the public – was one of the first women elected as a Fellow of the Royal Astronomical Society after the ban on women was removed in 1916. The Royal Astronomical Society now awards an annual Annie Maunder Medal to great space communicators of today.

building back up again. The amount of other solar activity – the solar flares, prominences and coronal mass ejections we'll see in the next sections – also follows this trend. It takes roughly eleven years of differential rotation to twist up the magnetic field of the sun sufficiently that it snaps, only to then reset and wind back up again.

Astronomers have picked up on other patterns, too. The first is called Spörer's law after the German astronomer Gustav Spörer. Early in the eleven-year cycle, sunspots appear at high or low solar latitudes – i.e. far from the sun's equator. However, as the cycle progress, they appear closer and closer to the equator. The positions of sunspots over time plotted on a graph looks like a butterfly – hence it is known as a Butterfly Diagram. Joy's law, named after American astronomer Alfred Joy, says that a pair of sunspots is often tilted, with the leading spot nearer to the solar equator.

Flares, prominences and filaments

Most people think that you can never look at the sun. While this is certainly good advice in most cases – it can blind you very quickly – you can look directly at the sun if you're using a specialized solar telescope. Large filters at the front knock down the intensity of the light, only letting a tiny, safe sliver through.

If you look at the sun in this way then it is very likely that along with sunspots you will see what appear to be tiny flames licking off the solar limb. These are *solar prominences*. When the sun's magnetic field lines burst outwards into space they carry some of the hot gas with them. Arguably they're at their most spectacular when they form an arch towering above the photosphere – the hot gas is following the magnetic field lines out of, and then back into, the sun. They may look small, but these arches often stretch for hundreds of thousands of kilometres.

Exactly what you see depends on the angle you're looking from. Imagine a prominence erupting straight out of the sun directly towards you. You'd see it head-on rather than edge-on. Astronomers call these events *filaments*. They look like snakes slithering across the solar disc. Like sunspots, they appear darker because you're seeing cooler gas in front of the hot surface of the sun beyond.

It's not uncommon for people to mistakenly call prominences solar flares, but flares are a separate class of solar phenomenon. As their name suggests, they involve a sudden brightening of a localized region of the sun and an outburst of radiation. The energies involved can be staggering – a single flare can release as much energy as a billion megatons of TNT. To put that into context, all the explosives used in the Second World War – including the atomic bombs dropped on Hiroshima and Nagasaki – totalled just three megatons of TNT.

Solar flares are often accompanied by the most spectacular outbursts the sun has to offer: a Coronal Mass Ejection (CME).

Coronal Mass Ejections

In March 1989, 6 million people in the Canadian province of Quebec were plunged into darkness during a nine-hour power blackout. Meanwhile, communications with weather satellites were interrupted and the Northern

Lights were no longer so northern – they could be seen as far south as Texas and Florida. All these events had the same root cause: a Coronal Mass Ejection.

These violent outbursts from the sun launch a billion tonnes of material into space at over a million kilometres per hour, flooding the solar system with charged particles. If they reach the Earth, we are hit by a geomagnetic storm causing disturbances in our magnetic field. This produces extra electric current which trips out power grids, cripples satellites and intensifies aurorae. The sun kicks out a CME every three to five days, but fortunately most miss our tiny planet.

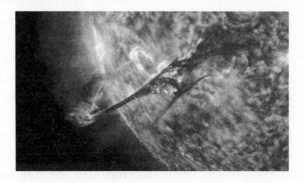

The sun unleashed a spectacular and powerful CME in August 2012.

One of the most spectacular CMEs to reach the Earth was the so-called Carrington Event of 1859, named after English astronomer Richard Carrington. Luckily, our electrical infrastructure was in its infancy at the time. The electrical telegraph was the most cutting-edge communication system around and it failed. Many telegraph operators reported receiving electric shocks. Should a similar event strike today, the damage would run into trillions of US dollars. Planes would have to be grounded while the storm subsided. Even now, pilots and aircrew are classed as radiation workers. During a much lesser solar storm in 2003, anyone on a flight from Chicago to Beijing would have been exposed to 12 per cent of their annual radiation limit.

Understandably, there is a desire to predict these events – to have a space weather forecast, much like we have a terrestrial one. We can't stop it, but we can minimize the damage. At the moment we can't tell if an approaching storm is the dangerous kind until a few hours before it hits. Some commentators have said that space weather forecasting is currently thirty to forty years behind its terrestrial cousin. But steps are being taken to push this notice window out beyond twenty-four hours, and then into days. This work is crucial – events equivalent to Carrington's in 1859 are thought to occur every 150 years or so. It's only a matter of time before another heads our way.

The solar wind

The drill was relentlessly rehearsed. The parachute would open and a waiting helicopter would spear the chute, bringing the intrepid traveller to a soft landing. Except it didn't quite happen like that. On 8 September 2004, NASA's *Genesis* probe tore down through the atmosphere and crashed straight into the ground. Photos of the crash site show the helicopters looking on, almost forlorn.

The drogue parachute failed to deploy – someone had installed the accelerometer upside down. Almost all of *Genesis*'s precious cargo was contaminated beyond use.

Luckily, however, some samples were recovered intact. The probe was launched three years earlier in a daring attempt to capture particles from the solar wind and return them to the Earth for analysis. It was the first sample return mission since the Apollo era, and the first ever to bring back material from beyond lunar orbit.

The idea of invisible particles streaming out from the sun had been mooted as far back as 1859 by Richard Carrington during the Carrington Event. Today, we know these charged particles – mostly electrons and protons – blow out from the sun in all directions at over a million kilometres an hour. The fastest winds rage away from holes in the sun's corona. The particles reach all the way out beyond the orbits of the planets, where they meet crosswinds from other stars (see page 117).

When the solar wind buffets the Earth's magnetic field, it induces aurorae near to the poles (see page 75). Yet the solar wind is far from serene – it also has the power to destroy. Astronomers think Mars's atmosphere was much thicker in the past, capable of sustaining liquid water on the Martian surface. Without a magnetic field, however, the solar wind has slowly eaten away at Mars's atmosphere, stripping it almost bare. Now Mars is a dry, barren wasteland (see page 91).

The Earth

Formation and structure

The Earth was fashioned from scraps and offcuts. The main event in this part of the universe was the formation of the sun some 4.6 billion years ago. But a substantial amount of gas and dust remained swirling around the infant star. Gravity slowly gathered this material into larger objects called *planetesimals* – the building blocks of planets – each roughly one kilometre across.

By 4.56 billion years ago, just a few hundred million years after the formation of the sun, some of these chunks of rock and metal had smashed together to create the Earth. Constant bombardment by fresh planetesimals, along with energy provided by radioactive decay, kept the newly formed planet molten. Gravity rounded out the resulting ball into a sphere.

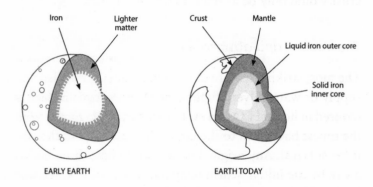

The early Earth was completely molten, allowing heavy materials like iron to sink towards its core.

With the planet molten, the heaviest material was free to sink to the bottom. The lightest material ended up floating to the top. Geologists call this process *differentiation*. As the differentiated Earth cooled, a solid crust formed above a dense iron and nickel core.

Today our planet still has an iron/nickel core. It is split into two parts – the inner and outer core. The inner core is solid, the outer core is molten due to the immense crushing force of all the material above. Temperatures at the boundary between the inner and outer core can reach 6,000 degrees Celsius, making it as hot as the surface of the sun. The inner and outer core make up the inner 55 per cent of the planet and are surrounded by the mantle which is made of semi-molten rock called magma. On top of the mantle sits the crust – the surface of the Earth on which we live. Just 60 kilometres deep at its thickest part, the crust contributes less than half a per cent to the Earth's diameter. Shrink our planet to the size of an apple and the crust would only be as thick as the fruit's skin.

The oceans and atmosphere

The most striking feature of our blue planet is the plentiful supply of water. Over 70 per cent of the Earth's surface is covered in liquid H_2O and every living thing on Earth, from the tiniest bacterium to the largest blue whale, depends on it for survival. Any original water would have been boiled away by the initial hellish temperatures, so the water was likely added to the surface later. But from where?

It may have been manufactured deep below the crust in the mantle. Liquid hydrogen and quartz can react to create liquid water that then becomes trapped inside rocks. A deep reservoir of water capable of filling the surface oceans three times over was discovered in 2014, 700 kilometres beneath the Earth's surface. Over time, water vapour probably oozed out of cracks in the crust. Then, as the planet cooled, the vapour condensed into liquid and rain filled low-lying basins.

Another possible source of water is outer space – delivered here by asteroids and comets as they crashed into our planet. But there are problems with this idea. Analysis of comets suggests many contain a different type of water to that found in our oceans (see page 99). If water came via asteroids instead, there should be a lot more xenon in our atmosphere than there is. So the jury is still out.

The origin of our atmosphere is a little clearer cut, but it started out far from its current composition. The first gases to cling to the infant Earth were released by volcanic activity from deep inside the planet. That means mostly carbon dioxide, laced with carbon monoxide, hydrogen sulphide and methane. There was no free oxygen. It was all locked up inside water (H_2O) and rocky, silicon compounds such as silicon dioxide (SiO_2).

Then everything changed around 3 billion years ago when microscopic organisms called cyanobacteria started to flourish in the oceans. They created free oxygen by combining carbon dioxide, water and sunlight through *photosynthesis*. The build-up of oxygen in the atmosphere led to one of the greatest mass extinctions in Earth's history as oxygen was toxic to the vast majority of life forms. Only organisms that could adapt to this massive shift in the composition of our atmosphere survived. You are descended from those survivors. Today, oxygen is the second most abundant element in our atmosphere (21 per cent) behind nitrogen (78 per cent).

Tectonic plates

The vast Himalayan mountain range buttressing the border between the Tibetan plateau and the Indian subcontinent is one of Earth's natural wonders. Every year thousands of people are drawn by the majesty of Mount Everest and hundreds attempt to scale the world's highest peak.

But, compared to the age of the Earth, the Himalayas are very young indeed. Most estimates place their formation at just 50 million years ago. The land that now largely falls within the borders of India travelled on a quite remarkable path to end up where it has. It broke off from an ancient continent known as Gondwana, before heading northwards to deposit Madagascar next to mainland Africa, then continued its journey towards Asia. Travelling at speeds of 20 centimetres per year, it ploughed headlong into the planet's biggest continent to create the world's tallest mountain range.

This sizeable movement of land mass is only possible because the Earth's crust is formed from a series of tectonic plates floating on a liquid ocean of molten rock. Roiling undercurrents caused the Indian plate to separate from Gondwana and head north. Upon meeting, the Indian plate was driven under the Eurasian plate, forcing material upwards and forging the Himalayas. This process is far from finished. The collision only slowed the progress of the Indian plate – it is still moving north. The Himalayas continue to rise by two centimetres every year.

But tectonic plates are not just a geological curiosity. Many scientists believe they played a key role in the development of life on Earth. After all, Earth is the only planet in the solar system to have them. Volcanoes often form at plate boundaries, allowing gases trapped beneath the planet's surface to escape into the atmosphere – carbon dioxide in particular. Extra carbon dioxide during ice ages helps raise the temperature. But plate movement can also trap excess CO_2, stopping the planet from overheating.

So, when it comes to searching for life elsewhere in the universe, astronomers are keen to find not only planets that are the same temperature as our own but also those with tectonic plates to keep that temperature within biology-friendly limits.

ALFRED WEGENER (1880–1930)

Look at the surface of the Earth and it resembles a giant jigsaw puzzle. The land that juts out from the top right of South America fits neatly into the gap underneath Western Africa. Noticing this, and believing it to be more than just coincidence, German physicist Alfred Wegener published his theory of continental drift in 1911, suggesting the two continents had once been one. It received little fanfare – other scientists couldn't believe that such huge masses of land could move and Wegener was unable to say why they moved.

It would take until the 1950s and 60s, long after Wegener's death during an expedition in Greenland, for supporting evidence to emerge. Scientists discovered that the seafloor spreads out over time as volcanic activity lays down new oceanic crust. The theory of plate tectonics soon followed, finally providing a mechanism to explain Wegener's original idea of continental drift.

Tides

At the north-eastern end of the Gulf of Maine, on the undulating Atlantic coastline of North America, sits a unique inlet called the Bay of Fundy. Twice a day, over 100 billion tonnes of water floods and empties from the bay. To put that into context, that's more water than flows through all of Earth's freshwater rivers combined.

What's the cause of such an enormous movement of water? Gravity. Specifically the gravitational pull of the moon (and partly the sun), which causes huge tides to rise and fall around the world each day. The Earth's rock is pulled on, too, but the water is much freer to move. The effect is most extreme in the Bay of Fundy where the tidal difference ranges from 3.5 metres to 16 metres – the water rises by more than the height of a four-storey house.

It works like this. When your part of the world is facing the moon, your local water is pulled towards it and

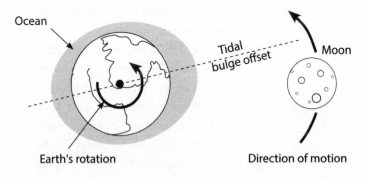

Ocean

Tidal bulge offset

Moon

Earth's rotation

Direction of motion

The moon's gravity creates a tidal bulge on the side of the Earth closest to it.

away from the regions of the globe at right angles to you. You experience a high tide and those regions experience low tide. The moon isn't pulling as hard on the water on the opposite side of the planet because it is further away. Here there's another high tide due to the centrifugal force of a spinning Earth – the same force that pushes you outwards into the side of a car when taking a sharp bend. That's why most places on Earth experience two high tides and two low tides each day – the spin of the planet carries us through those four regions in each twenty-four-hour period.

The reality of the situation is worth thinking about for a moment. If you've been at the beach and seen the tide go out then it's likely you won't take much convincing that the water is moving away from you. But that's not really what's going on. The water is mostly staying where it is, held in place either by the pull of the moon or by centrifugal forces. Instead, it is you that is moving – you're being carried out of a tidal bulge by the rotation of the Earth. You and the beach are running away from the water!

The seasons

One of the most beautiful features of our planet is the seasonal variation. In spring, flowers burst from the ground and reach for the sky. By autumn, leaves are falling in the opposite direction. A lot of people make the mistake of thinking the changing temperatures throughout the year are due to our distance from the sun. That perhaps we're closer in the summer and further away in winter.

Instead our seasons are the result of the tilt of the Earth. Our planet doesn't sit bolt upright – it is tipped over from the vertical by 23.4 degrees. This means that in June the northern hemisphere is tipped in towards the sun and people there see warmer weather and longer days. Those living inside the Arctic Circle experience perpetual day – they're so tilted towards the sun that it never sets. Meanwhile, the southern hemisphere is leaning away from the sun and so heat and light are harder to come by – it is winter. Antarctica is plunged into permanent night.

Six months later, with the Earth on the other side of the sun, the roles are reversed. Those beneath the equator are dusting off the barbecue while many above are reaching for a jumper. The Arctic is in shadow and Antarctica is constantly illuminated.

The longest and shortest days of the year (in June and December) are called solstices. Immediately in between these dates we reach a point where neither hemisphere is tipped in nor away from the sun. On these days in March

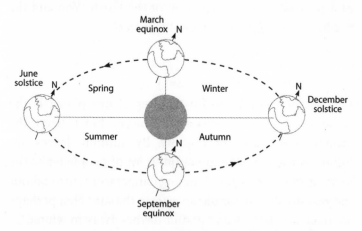

We experience seasons because Earth's tilt means we're tipped towards the sun at some points and leaning away at others.

and September – known as equinoxes – the whole planet experiences equal hours of daylight and night.

We should be thankful that our planet's tilt is reasonably small. Were the angle more severe, our seasonal variation would be a lot more dramatic and harder to cope with. Thanks to the moon, our tilt also remains stable and the seasons predictable. The axis of Mars, for example, without a large moon for ballast, swings wildly under the gravitational pull of the other planets. This results in long winters and intense summers that are unequal and constantly shifting.

Magnetic field

The journeys undertaken by female sea turtles are extraordinary. Born on the beach, they scurry down to the sea and migrate more than 2,000 kilometres to find rich feeding grounds. Yet once mature they return all the way back to the very same beach they hatched on. How do they remember where they came from? The answer, it seems, is that they use the Earth's magnetic field.

Deep in the bowels of our planet, molten iron sloshes around in the outer core as the Earth spins. This motion generates the Earth's magnetic field. It bursts out of the top of the planet, looping back round into the opposite end. However, when it comes to poles, things get a little complicated. The Earth has three north poles and three south poles.

First, there is the Geographic North Pole – the physical point at the top of the planet that coincides with the imaginary line of the Earth's rotation axis. It is pretty much fixed, only moving a few metres a year in a repeating cycle. Then there's the North Magnetic Pole – the place a compass points to. A compass with a needle free to move vertically would point straight into the ground at this point. Due to changes in the Earth's outer core, the North Magnetic Pole is highly mobile. Until recently it was located in Canada, but it's now wandering across the Arctic towards Siberia. The geographical poles and magnetic poles are currently misaligned by approximately ten degrees.

Finally, there's the Geomagnetic North Pole. This is where the magnetic north pole would be if you placed an ordinary bar magnet in the centre of the Earth. In reality, our magnetic field is a lot more complicated than a bar magnet's. These three northern poles all have equivalents in the southern hemisphere.

Without this magnetic field it is hard to imagine life thriving on Earth. It acts as a giant force field, deflecting harmful radiation from the sun and deep space. It buffers us from the solar wind, which would otherwise strip away our ozone layer and leave us susceptible to an increase in ultraviolet rays from the sun.

Aurorae

Appearing as vast curtains of green light shimmering and pirouetting in multiple directions across the sky, the Northern Lights (or *aurora borealis*) can regularly be seen close to the North Pole. The same effect is seen close to the South Pole, too (*aurora australis*). They also make a surprising array of sounds, including hisses, pops, crackles and muffled bangs.

They are a stunning reminder that our planet is not isolated in space, but rather it has an intimate relationship with its star. We are constantly bathed in the solar wind – the gust of charged particles blowing out from the sun. Interactions between the solar wind and the Earth's magnetic field cause charged particles to accelerate along the planet's magnetic field lines towards the poles. There they slam into the atmosphere above our heads, providing the atoms in the air with additional energy. We see aurorae when those atoms release that new-found energy in the form of light.

The effect is normally restricted to a limited area around each magnetic pole known as the auroral ovals. However, a geomagnetic storm – such as those caused by a Coronal Mass Ejection – can overwhelm our magnetic field and widen the ovals considerably. During the 1859 Carrington Event, sailors in the Caribbean reported seeing fantastic light displays. Having never sailed near the poles they had no idea what they were. The glow was so bright over the Rocky Mountains that miners awoke thinking it was daytime. Even those in Sub-Saharan Africa were treated to the show.

Green is the most common hue because that's the colour given out by oxygen at low altitudes where it is most abundant and most easily visible. Red bands of auroral light represent light emitted by oxygen under quieter circumstances much higher up in the atmosphere. If you can spot blue streaks, they are caused by interactions with nitrogen.

Earth is not the only planet in the solar system to experience aurorae. Astronomers have spotted the same effect on Mars, Jupiter and Saturn.

Meteorites and shooting stars

Several million years ago, an impact on Mars splintered part of the Martian surface and ejected it into space. Somehow it managed to cover the 225 million kilometres across the void to the Earth, crash down through the atmosphere and embed itself in the Antarctic tundra as a meteorite. In the world's collection of meteorites these Martian interlopers are exceptionally rare – making up less than half a per cent of the total. A little more common are meteorites from the moon. However, the vast majority of meteorites come from the sun's family of asteroids – lumps of rock and metal left over from the formation of the solar system. And that's their great attraction – most pre-date the Earth itself, so bring with them valuable clues about how the sun and its orbiting worlds came to be.

What you call a piece of space debris depends on where it is. A small lump of rock still in space is called a meteoroid. As it tears through an atmosphere the term switches to meteor. Only if it makes it to the surface of a world intact does it get labelled a meteorite.

Many meteors at once create a dazzling spectacle called a meteor shower – the sudden onset of a bevy of 'shooting stars'. During its journey around the sun, the Earth regularly ploughs through trails of space dust strewn around the solar system by passing comets. As these tiny particles – often no bigger than a grain of sand – glow incandescent through friction with the atmosphere, they are seen to streak across the sky.

One of the most spectacular showers is the Perseids in August. From a dark site, far from intrusive city lights, you'll spot a meteor a minute lighting up the night sky. It's an annual reminder that there's a lot more to the solar system than just the planets.

Satellites and the International Space Station

The 4th of October 1957 was a landmark day in human history. The Soviet satellite *Sputnik 1* became the first artificial object to orbit the Earth. Three months later it re-entered the atmosphere and burned up. Since then, satellites have transformed the way we live. Weather stations keep track of the climate, spy satellites surveil enemies, TV satellites bring us the latest must-watch shows and the Global Positioning System (GPS) helps us find our way.

There are over a thousand operational satellites up there, but not everything in orbit is useful. Right now there are over 21,000 objects larger than ten centimetres across whizzing around our planet. For objects with diameters between one and ten centimetres the number rises to half a million. Most of this is space junk – debris from satellites and space missions left to float in a growing swarm of rubbish.

This is a problem for the biggest artificial satellite of the Earth – the International Space Station (ISS). Roughly the size of a football field, it is home to a crew of six astronauts from across the globe. It orbits at around 400 kilometres above the Earth, but that orbit has been slightly altered several times to dodge large pieces of space junk. The outer hull, particularly the solar panels, bears the scars from impacts with smaller debris.

Astronauts swap in and out, usually every six months, with the ISS permanently inhabited since 2000. Taking just ninety-two minutes to orbit the Earth, the crew is treated to sixteen sunsets and sixteen sunrises each day. They also get spectacular views of the Earth from above, including our sprawling cities, powerful thunderstorms and a front-row seat for dancing aurorae.

As well as being a beacon of international cooperation, it is designed to extend our knowledge of the effect of long stays in space on the human body. Ultimately, we'll use the lessons we learn there to send people to Mars.

The moon

Formation

Our moon is an oddball. When you compare all the moons in the solar system to their planets, the Earth–moon relationship is way out in front when it comes to size. The moon's diameter is nearly 28 per cent that of Earth's. The next nearest is Neptune's largest moon, Triton, which is only 5 per cent as wide as its host. We have the fifth-largest moon in the solar system, but we're only the fifth-largest planet.

That means it is very unlikely the moon formed elsewhere and was gravitationally captured by us at some later date. It is just too big. Charles Darwin's son George believed the moon was spun off from the Earth, leaving behind the land-free gap that's now the Pacific Ocean.

Today, the leading idea for its origin is that Earth was blindsided in its youth by a planet about the size of Mars. Astronomers call this lost world Theia and the calamitous collision the Giant Impact Hypothesis. It is thought to have occurred just 50–100 million years after the Earth's formation. The crash sent a ring of debris spiralling into orbit and gravity eventually drew the material together to form the moon. If Earth's history is condensed into a single day, the moon formed when it was just ten minutes old.

Such an event explains why the moon has an uncharacteristically small core and why it is less dense than the Earth. The heaviest material from Theia stayed close to the Earth, the lighter stuff thrown outwards to coalesce into the moon. The Earth has the highest density of any planet in the solar system, which makes sense if additional material from Theia was added to it after its formation. A giant impact also accounts for why the moon appears to have been molten in the past – the violent collisions between material in the ring liquefied the rock.

Further evidence comes from the moon rocks returned by the Apollo missions. Computer simulations suggest that the moon was formed mostly of material from Theia, so there should be differences between terrestrial and lunar rocks. In 2014, scientists announced they'd found a slightly different mix of oxygen types in the Apollo samples.

Craters, seas and phases

Look at a map of the moon and you'll find it's full of poetic-sounding places such as the Sea of Dreams, the Bay of Rainbows and the Lake of Happiness. In reality it is a very hostile environment with almost no atmosphere. The moon's tenuous gas layers weigh less than five elephants.

Its distinctive surface is covered in sprawling dark patches called *maria* or seas (singular *mare* – 'mah-ray'). Seen from the Earth they look a bit like a face, hence the famous Man in the Moon. However, they are not seas of water and never have been. Instead they were vast vats of molten lava formed during the moon's tumultuous birth that have since cooled and solidified. The seas are pockmarked with thousands of craters – deep, bowl-shaped scars from impacts that have peppered the lunar surface over billions of years.

The moon's appearance in our sky changes regularly. That's because it doesn't make any light of its own. Instead it acts as a giant mirror, reflecting sunlight towards us. How much reflected light we see depends on where the moon is in its orbit around the Earth. When it sits between us and the sun, all the light strikes the far side of the moon and bounces away from us. We call this a *new moon*. As the moon heads for the opposite side of the planet, we gradually see more of it illuminated until, around two weeks later, it reflects back to the Earth all the light hitting it. We see a *full moon*. We begin to lose that light on its return journey as less and less falls on the side facing the Earth.

Tidal locking

The fact that we only ever see one face of the moon naturally leads many people to think that it mustn't rotate. But it does. The moon spins on its own axis in exactly the same time as it takes to make one orbit of the Earth – 27.3 days.

To see how this works, find something to act as the Earth and place it on the floor. Stand facing it, then move around it in a circle always looking inwards. Once you get back to where you started you've not only made one lap of the Earth, you've also spun once on the spot. To convince yourself of this, repeat the exercise but focus on the walls you're looking at as you move round. You'll see that you face each of the four walls of the room in turn, just as you would if you were spinning on the spot.

The moon behaves in this way because it is tidally locked to the Earth. The moon originally rotated much faster than it orbited our planet. But Earth's gravity stretched the moon slightly in a line between the two bodies. This made the moon slightly fatter in one direction than the other – a tidal bulge. The Earth then pulled preferentially on this bulge, gradually slowing the moon's rotation period until it matched its orbital period.

The moon's series of phases takes slightly longer to complete than its 27.3-day orbital period. The gap between full moons is 29.5 days. That's because the Earth, moon and sun must all be in a line for us to see a full moon. As the moon is busy going around us, we are also orbiting the sun. It takes a couple of days for the moon to make up the extra distance we've moved around the sun in a month and get back in line.

Tidal locking is a common feature in the universe. Many of the moons of Jupiter and Saturn are locked to their planet. Some planets in other solar systems are also tidally locked to their host stars. The possibility of life on such worlds – with one side baked in heat, the other shivering in darkness – is hotly debated among astronomers (see page 152).

Importance to life on Earth

The moon comes wrapped in so many cultural stories, often dating back thousands of years, that it is difficult to separate the scientific reality from the old wives' tales. Ideas that the moon directly affects human behaviour can be seen in reports of werewolves and lunatics – people who are literally moonstruck. Midwives will swear their maternity wards are more crowded during a full moon. Yet there's no rigorous proof to back up these claims.

Most of these explanations invoke the moon's gravitational pull, arguing it is stronger at full moon and that this has an effect on the water in our bodies. But the moon's closest approach to the Earth rarely coincides with a full moon – it can just as easily be closest at new moon. With those caveats in mind, the moon has hugely influenced life on Earth. Many scientists believe that without the guiding hand of our nearest neighbour we wouldn't be here to marvel at it.

We've already seen how it stabilizes the seasons (see page 74). It might also have played a starring role in kick-starting life on this planet. Forming about fifteen times closer to the Earth as it is now, its gravity raised monster tides which encroached hundreds of kilometres inland with a much greater frequency than today's tides. Some researchers believe life got started in these tidal areas, where the churning of land and sea mixed the building blocks of biology into a form capable of developing into life.

The moon is also slowing our planet down. A billion years ago, the day was eighteen hours long. It now lasts twenty-four hours because the Earth is gradually losing rotational energy through friction with the oceans as it tries to turn underneath water held in place by the moon's gravity (see picture on page 71). This transfer of energy to the oceans helps transport heat from the equator to the poles, making the temperature range across the Earth considerably smaller. So, once life got started, the moon helped keep conditions favourable for it to evolve and diversify.

A consequence of the Earth slowing down is that the moon is moving away from us, meaning the tides are calmer than they used to be – another reason why conditions here are now more stable. We can measure how quickly the moon is fleeing thanks to experiments left on the moon by the Apollo astronauts.

The Apollo missions

'Contact light. OK, engine stop.' These ordinary words ushered in an extraordinary period in human history. Thirty-eight-year-old Neil Armstrong had just manually landed the *Eagle* on the moon after a hair-raising descent over a field of sizeable boulders. They had less than a minute's worth of fuel left. Mission control was understandably relieved. 'We copy you on the ground. You got a bunch of guys about to turn blue. We're breathing again.'

A few hours later, Armstrong shuffled down the ladder and became the first human being to set foot on another world. That day, 20 July 1969, remains a beacon of what can be achieved when we set our minds to it. Over the next three years, NASA successfully landed five more missions and ten more men on the moon. Only *Apollo 13* had to be abandoned after an exploding fuel tank crippled the craft mid-flight.

Edwin 'Buzz' Aldrin on the moon during the *Apollo 11* mission. The photographer, Neil Armstrong, can be seen reflected in his visor.

The missions weren't just about one-upping the Soviet Union at the height of the Cold War. They were also scientifically valuable. The six missions returned a total of 382 kg of moon rocks that have revealed important clues as to how our nearest neighbour formed. Banks of mirrors were left on the surface so we can fire lasers from Earth to keep track of how fast the moon is running away from us (currently 3.8 centimetres per year). Moonquake detectors were embedded in the dust to study lunar tremors.

The later missions got more daring, taking buggies to the moon to drive around and explore more of the barren landscape. Alan Shepard even smuggled the head of a 6-iron golf club aboard and hit a shot on the moon.

Dave Scott dropped a hammer and a feather to show that objects of different masses fall at the same rate without the braking effect of an atmosphere.

As *Apollo 17* departed on 14 December 1972, its commander Gene Cernan – the last man on the moon – was hopeful of a return. Objections to the cost mean we've never been back. But the pull of the moon is irresistible. It's a natural place to extend long stays in space and various space agencies around the world are already working on plans to return. One day we'll leave our footprints in the lunar dust again.

The Late Heavy Bombardment

It appears that the inner solar system was carpet-bombed 3.9 billion years ago. Long after the initial chaos of the solar system's formation, there was a sudden spike in the number of impacts raining down on the rocky planets. While Earth's bruises from this event have long since faded thanks to erosion, the airless moon still bears the scars.

Due to the fact that it happened 600 million years after the solar system's birth, and was particularly ferocious, astronomers call this event the Late Heavy Bombardment. The leading explanation points the finger at Jupiter. Computer simulations of solar system formation suggest that the giant planets were unlikely to have formed in their current locations (see page 120). It seems Jupiter moved inwards towards the sun. This would have scattered asteroids like a flock of pigeons. Many would have smacked into the moon and the rocky planets.

But not everyone is convinced. The main evidence for

a Late Heavy Bombardment comes from moon rocks returned by the Apollo missions. Rocks from multiple sites across the lunar surface all point to collisions around the same time. However, some astronomers have argued that just a few large impacts could have catapulted debris far and wide, contaminating multiple sites. That would make a trickle of impacts look like a deluge.

The other problem is the emergence of life on Earth. The traditional picture has the infant Earth as a fiery hellscape too hot and too relentlessly battered for life to gain a foothold. In this picture, life could only get started after the Late Heavy Bombardment. And yet recent evidence is pointing to an Earth that already had oceans of water and perhaps even life as early as 4.1 billion years ago.

So either life survived the onslaught, got wiped out only to re-emerge, or the Late Heavy Bombardment didn't happen as we've traditionally thought. Whichever it is, the period is the focus of a lot of current research into the solar system's tumultuous past.

CHAPTER 3

The Solar System

Mercury

It's a bare, rocky place. In fact, at first glance, you might mistake Mercury for the moon. The nearest planet to the sun is baked during the day, with temperatures rising to over 400 degrees Celsius. Without an atmosphere to trap the heat, however, night-time temperatures plummet to around –200 degrees Celsius. It takes Mercury – the smallest planet in the solar system – just eighty-eight days to orbit the sun. One Mercurian day lasts nearly fifty-nine Earth days.

Only two spacecraft have visited Mercury. The first – *Mariner 10* – flew by in the mid-1970s. Then *MESSENGER* went into orbit there in 2011. It circled Mercury over four thousand times before scientists deliberately crashed it into the planet in April 2015. It's likely that *MESSENGER* is the only thing ever to have orbited Mercury, human-made or otherwise. The close proximity to the sun, with its intense gravitational pull, prevents any moons from forming around the planet or being captured from elsewhere. That gravitational pull also means Mercury spins on its axis three times for every two orbits it makes.

These missions provided rich details of a world difficult to observe from the Earth due to its diminutive size – it is smaller than some of the moons of Jupiter and Saturn. Mercury's most distinctive feature is the Caloris Basin – an ancient impact crater that ranks as one of the largest in the solar system. Over 1,500 kilometres across, it was discovered during the fly-by of *Mariner 10*. The ground is ruffled up on the exact opposite side of the basin, the terrain becoming hilly and grooved. It's thought the impact rang Mercury like a bell, sending shock waves hurtling in opposite directions around the planet. The grooved terrain is the result of the shock waves colliding exactly 180 degrees away from the point of impact.

Like Venus, we also see Mercury occasionally pass in front of the sun. However, Mercury transits are far more common. We get thirteen or fourteen a century. In a lovely twist, on 3 June 2014 the *Curiosity* rover on Mars watched Mercury ghost in front of the sun. This transit wasn't visible from the Earth, and it marked the first observation of a transit of either Mercury or Venus from the surface of another planet.

Venus

There's no getting away from it: Venus is a horrid place. It's enshrouded in thick layers of carbon dioxide laced with sulphuric acid. This heaving atmosphere traps immense amounts of heat from the sun and presses down on the surface with an atmospheric pressure ninety-three times that of Earth's. Anyone foolish enough to venture down there would be simultaneously baked, crushed and dissolved.

Despite not being the closest planet to the sun, these cloud decks make Venus the solar system's hottest planet. Thanks to an intense greenhouse effect, temperatures there can rise nearly 40 degrees higher than on Mercury. However, these extreme conditions didn't stop the Soviet Union from landing several probes on Venus, starting with *Venera 9* in 1975. It was the first spacecraft to return a photograph from the surface of another planet. It managed to hold out for just fifty-three minutes before succumbing to Venus's hellish environment.

Venus is the closest planet to us, and is often called 'Earth's twin', but the only real similarity is size – Venus is 95 per cent the diameter of the Earth. The planet's quirkiest feature is that its day is longer than its year. That may sound utterly perverse given that we live on a planet where days are significantly shorter than years. Yet Venus's slow rotation means it takes 243 Earth days to spin once on its axis. It only takes 225 days to orbit the sun. It is also the only planet to spin clockwise. It's very unlikely Venus started out like this. Perhaps a collision with a large object knocked it over to spin upside down with a much slower rotation rate.

In more recent years, Venus has been visited by the *Magellan* and *Venus Express* missions. *Magellan* produced glorious radar maps of the surface, allowing us to peer deep below the clouds. Conspicuous features include Skadi Mons and Maat Mons – Venus's two tallest mountains. The former forms part of the Maxwell Montes mountain range, named after Scottish physicist James Clerk Maxwell.

It remains the only feature on Venus not named after a woman or goddess.

It is also worth noting the adjective that astronomers have assigned to Venus. Really we should refer to its features as Venerean, much as we'd say Mercurian or Martian. However, you can perhaps see why that's avoided. Astronomers tend to use Venusian as a more family-friendly alternative.

Mars

Of all the planets, Mars has captivated and enthralled us like no other. Throughout human history people have made sacrifices to it, been fearful of invasions from it and sent car-sized rovers to explore it. It remains the planet we know the most about. We even have a better map of the Martian surface than we do of the ocean floor here on Earth.

Its distinctive orangey-brown hue, which gives it the often-used epithet of the Red Planet, is due to high levels of iron oxide (rust) in the rocks there. In the night sky you can see that Mars has a ruddy appearance, even without a telescope. Today it is a dry, cold desert of a planet. Yet perhaps it wasn't always this way. Our missions to Mars have found clues that it could have been a very different world in the past, perhaps with oceans of water covering up to a third of its surface.

Why Mars's climate changed so dramatically is still very much an open question. The leading idea is that the planet's core solidified over time because, as a smaller planet, there isn't as much gravitational pressure from above. That would also have seen Mars's magnetic field switch off, leaving it unprotected from the ravages of the solar wind. The Red Planet's atmosphere was severely depleted over time. Now it is only blanketed by a thin wisp of carbon dioxide. Mars's atmospheric pressure is now so puny – less than 1 per cent of Earth's – that ice skips the liquid phase entirely, jumping straight to vapour in a process called *sublimation*.

Mars has a split personality: above the equator it is incredibly flat, whereas below is dominated by mountains. Their only similarity is poles both capped with ice. Mars's southern hemisphere is home to Olympus Mons, the solar system's tallest volcano and second-highest peak. It is more than twice the height of Mount Everest, but would be considerably easier to climb. The sides of the volcano slope at just five degrees. Don't expect to see the top from the bottom though – the volcano is so wide that the peak would be hidden over the horizon.

There is also a huge, splintering valley of intricate canyons stretching almost a quarter of the way around the planet's equator. Called the Mariner Valley, it's part of a vast volcanic plateau known as the Tharsis bulge.

The planet is orbited by two moons – Phobos and Deimos. Their names mean fear and terror and are named after the sons of the god of war who accompanied him into battle. They are tiny, just 22.2 and 12.6 kilometres across respectively.

Robotic exploration

We've sent an armada of probes to orbit round, land on and drive across the Red Planet. They've witnessed sunsets on another world, seen dust devils swirl across Mars and even spotted our own planet in the sky of another. They are a remarkable testimony to the human spirit of discovery and exploration.

The main draw of these missions, from *Mariner 4* in 1965 to the more recent *Curiosity* rover, is to work out whether conditions on Mars have ever been favourable for life. The *Viking* landers of the 1970s – the first missions successfully to operate from the surface – carried experiments capable of directly testing the Martian soil for signs of biology. Initial results were positive, but the consensus of opinion is now that this was a false positive. These machines were restricted to the area immediately around their landing sites, but later missions sent wheeled robots to the surface to rove about.

The *Spirit* and *Opportunity* rovers, in particular, were a striking success. They touched down in 2004 and were designed to last just ninety days. *Spirit* managed to eke out six years and clock up nearly eight kilometres before it found itself marooned in a bed of soft soil. At the time of writing, *Opportunity* is still going strong, having covered a distance greater than an Olympic marathon.

They were joined on Mars by *Curiosity* in 2012. The size of a small car, it couldn't land in the same way as *Spirit* and *Opportunity*. They'd been deployed inside an inflatable cocoon that bounced several times before coming to a stop. *Curiosity* was lowered onto the surface using a futuristic-looking sky crane. It's well worth searching out the video of this white-knuckle ride to the Martian surface – it was a staggering feat of imagination and engineering.

HUMANS TO MARS

Humans could well go to Mars this century, but it's significantly harder than going to the moon, which is only a three-day, 380,000-kilometre journey away. Mars is a seven-month, 225-million-kilometre trek.

Long stays in space come loaded with the dangers of radiation. High-energy particles penetrating the skin dump their energy in your cells and damage DNA. Cancer, radiation sickness and cataracts can result and a high dose can be lethal. So astronauts need shielding in a way that's also lightweight enough not to hamper the affordability of the mission. Weight is a huge factor, too. Humans require food, water and oxygen. Getting this heavy payload to Mars is expensive; landing it there is dangerous. The Red Planet's thin atmosphere means there isn't much gas there to act as a brake.

Then there's the return journey. Robots don't care about coming home again. Humans probably do. You need to take enough fuel with you – or manufacture it from resources already there – to allow you to come home.

The asteroid belt

An international team of scientists are cruising at 39,000 feet aboard a dedicated flying laboratory. All eyes and instruments are trained on an object ripping through the atmosphere at over twelve kilometres per second. Meanwhile, on the ground, four teams of scouts spread out across a 20 by 200-kilometre strip of barren Australian outback, waiting for it to hit the ground. Eventually, they track it down, carefully package it up and take it away for further analysis.

Their quarry may have just come down from space, but it wasn't made there. The Japanese Space Agency (JAXA)'s *Hayabusa* probe had returned to the Earth after a glitchy seven-year expedition to the asteroid 25143 Itokawa. It was the first sample return mission from an asteroid.

Astronomers went to such great lengths because asteroids represent a unique opportunity to learn more about the solar system before it had planets. They are fossils from the earliest days of the solar system – planetary building blocks that didn't become part of a planet.

These lumps of tumbling rock and metal are found all over the solar system, but around 90 per cent form a belt-shaped collective between the orbits of Mars and Jupiter. The main asteroid belt represents part of a failed planet – one that couldn't form because of the disruptive gravity of neighbouring Jupiter.

The total mass of the belt is now only 4 per cent that of the moon. Just four asteroids – Ceres, Pallas, Vesta and Hygiea – contribute half of this mass. The rest get progressively smaller, until you reach pebbles and even dust particles. The larger asteroids have been the subject of much scrutiny. NASA's *Dawn* spacecraft visited Vesta in 2011, before departing a year later for Ceres. Arriving there in 2015, it became the first spacecraft in history to have orbited two different solar system bodies.

The main belt is home to around 2 million asteroids greater than one kilometre across. So you might think travelling through it is a treacherous, white-knuckle ride. Movies like *Stars Wars*, which see characters duck and dodge oncoming space rocks, help cement this idea. But space is huge. Illustrations or animations of the asteroid belt often blow up the size of the rocks many times over so you can see them. In reality, the average gap between asteroids is nearly a million kilometres.

Threat to Earth

The real danger of asteroids comes when they strike the Earth. Sixty-six million years ago, a 10-kilometre-wide asteroid – the size of a small city – careered into the Mexican coast, unleashing nothing short of hell. We can still see the crater it formed. Tsunamis tore through the oceans, fire rained from the sky, entire forests were flattened as carnage and mayhem ensued. Vast amounts of debris and dust thrown into the atmosphere plunged the Earth into a nuclear winter. Starved of sunlight, plants began to die. Then the creatures that ate plants died. Then the meat-eaters perished. Within 100 years, every dinosaur and 70 per cent of all land species had been extinguished. In the oceans it was as high as 90 per cent.

Thankfully, these extinction-level events are rare. Space rocks above five kilometres wide are only thought to strike the Earth once every 20 million years. And we do have one massive advantage over the dinosaurs: telescopes. Robotic telescopes are currently scanning the skies for objects above one kilometre in diameter and predicting their orbits for the next hundred years. The good news is there's nothing big coming our way any time soon.

However, that doesn't stop us getting blindsided by much smaller objects. In 2013, a fireball blazed through the skies above Chelyabinsk in Russia. A 20-metre-wide asteroid had snuck up on us, coming out of the sun like a Second World War fighter pilot springing a surprise attack. Fortunately no one was killed, but onlookers sustained injuries as the shock wave shattered windows and sent glass flying.

There will come a time when a large asteroid threatens Earth again. By then we should be able to do something about it. Unlike in Hollywood movies, nuking it is the worst possible option because you would just shatter it into slightly smaller chunks still heading for Earth. One of the best solutions is to keep it in one piece and gradually move it off course using the gravitational pull of a small space probe.

Planet	Diameter	Distance from Sun	Length of Day	Length of Year	Average Temp (°C)	Known Moons
Mercury	0.38	0.39	58.7 Earth days	88 Earth days	67	0
Venus	0.95	0.73	243 Earth days	225 Earth days	462	0
Earth	1	1	24 hours	365 days	15	1
Mars	0.53	1.52	24.6 hours	1.88 Earth years	-63	2
Jupiter	11.21	5.2	9.84 hours	11.86 Earth years	-161	69
Saturn	9.45	9.54	10.2 hours	29.46 Earth years	-189	62
Uranus	4	19.18	17.9 hours	84.07 Earth years	-220	27
Neptune	3.88	30.06	19.1 hours	164.81 Earth years	-218	14

Comet 67P, Rosetta and Philae

It was one of the most audacious feats in the history of robotic space exploration. After a ten-year, 6.4-billion-kilometre journey, the European Space Agency's *Rosetta* probe finally caught up with the comet known as 67P/Churyumov–Gerasimenko (or simply 67P). At the time it was between the orbits of Mars and Jupiter.

Along with asteroids, comets are denizens of the gaps between the planets. Unlike their rocky, metallic counterparts, they are largely made of ice. They also have highly elliptical orbits that carry them from the space out beyond Neptune to get up close and personal with the sun. Humans have wondered at comets for millennia due to the spectacle they put on as they pass our planet. Heated by the sun, and blasted by the solar wind, comets sport two tails that can trail for hundreds of millions of kilometres behind them.

ESA made history when their *Philae* lander touched down on the comet 67P in 2014.

The European Space Agency had sent a spacecraft to photograph a comet before, but they'd never attempted to land on one. Doing so was no mean feat given that 67P was tumbling around the sun at 55,000 kilometres per hour. Scientists watched on with trepidation as *Rosetta* despatched the washing-machine-sized *Philae* lander for a rendezvous with the surface on 12 November 2014.

It didn't go perfectly to plan. Harpoons designed to anchor *Philae* to the comet failed. The probe hit the surface and bounced several times, before coming to rest a kilometre away in the shadow of an ice cliff. Unable to use its solar panels in the darkness, the probe ran out of juice in two days. After more than six months of enforced hibernation, remarkably *Philae* awoke and hailed *Rosetta* again in June 2015. The comet's passage towards the sun had warmed the ice enough to free *Philae* from the shadows.

Much of the data from the mission is still being analysed by scientists, but one notable result is that the water on the comet seems different to that found on Earth. It contains a higher proportion of deuterium, which goes against the idea that comets similar to 67P helped deliver water to the early Earth.

Jupiter

The King of the Planets is a majestic sight, even through a small backyard telescope. You'd see its distinctive orange colour, along with layers of horizontal cloud belts. A slightly bigger telescope might also reveal its famous Great Red Spot. Although, be warned, Jupiter's rapid rotation – a Jovian day is less than ten hours long – might mean it is on the opposite side of the planet. The swiftness of its spin also means Jupiter is noticeably fatter at the equator than the poles.

The biggest of all worlds orbiting the sun, Jupiter could swallow up all the other planets with room to spare. You would need 1,321 Earths to match its volume. At an average distance of 778 million kilometres from the sun, it takes Jupiter nearly twelve years to complete one orbit.

It has roughly the same composition as the sun – 75 per cent hydrogen and 24 per cent helium. However, there is still a considerable amount of uncertainty about what's going on beneath the Jovian surface. There is thought to be a dense core at the centre, but we don't know how big it is. Astronomers also believe a layer of liquid hydrogen exists between the core and the outer atmosphere.

The cloud belts in that atmosphere often move in opposite directions. The dark areas are called zones and the lighter areas belts. Flashes of lightning a thousand times more powerful than Earth's have been detected there. The Great Red Spot, nestled in the cloud belts of Jupiter's southern hemisphere, is a long-lived anticyclone of staggering size. At one stage it was as big as four Earths. However, recent observations have revealed it's shrinking. We don't fully understand why, although small swirls of gas called eddies have been seen entering the storm and they may be changing its internal structure.

A little-known fact about Jupiter is that it has rings. In fact, all four of the giant planets have a ring system. Unlike Saturn's icy rings, Jupiter's are made of dust. They were only discovered when the *Voyager 1* probe flew by the planet in 1979.

As the solar system's biggest planet, Jupiter has the strongest gravitational pull. There is a lot of open debate about the role it's had in shaping our solar system. It has been implicated in the Late Heavy Bombardment, an intense barrage of impacts in the inner solar system (see page 86). However, it's unclear whether Jupiter is a friend or foe – whether it helps keep us safe by sweeping up impactors or threatens us by corralling them into dangerous positions. It's probably both.

Jupiter's moons

As you might expect, the biggest planet has the highest number of moons – sixty-nine at the last count. Most of these satellites are tiny – asteroids or comets that wandered too close to the giant world and got caught in its gravitational web. Some of these Jovian moons deserve just as much attention as the planets themselves. In particular, the four so-called Galilean moons discovered by Galileo Galilei in 1610: Io, Europa, Ganymede and Callisto.

Ganymede is the largest satellite in the solar system. At over 5,000 kilometres across, it's even bigger than Mercury. However, as we'll see, to be classified as a planet you need to directly orbit the sun (see page 112). Its neighbour Callisto has an ancient surface, little changed in the last 4 billion years. Its battered facade bears more impact scars that any other body in the solar system.

Arguably, however, the two most intriguing of Jupiter's moons are the innermost Galilean moons: Io and Europa. Io takes just 1.5 days to journey around the planet. Such proximity to Jupiter raises huge tides on the surface, expanding and contracting the small moon. This constant flexing – called *tidal heating* – melts the rock there and powers over 400 active volcanoes, making Io the solar system's most volcanically active place. Huge plumes of sulphur rocket hundreds of kilometres into the sky.

JUNO

In recent years NASA's *Juno* spacecraft has sent back photos of Jupiter in unprecedented detail. It arrived there in 2016, becoming only the second human-made object to orbit the solar system's biggest planet. The first – *Galileo* – had ended its mission in 2003. A decade of development in camera technology is more than evident in the wealth of spectacular images that flooded back to Earth.

Juno's close-up pictures of the Great Red Spot, taken as it swooped down over the planet, should help astronomers figure out why the famous feature is shrinking. Precise measurements of the gravitational pull the probe feels from the planet will help decipher what's going on in Jupiter's core. Understanding its atmospheric composition is key to understanding how it and the rest of the solar system formed.

A little-known fact about *Juno* is that it carried three aluminium Lego figures along for the ride. They represent the Roman god Jupiter, his wife – Juno – and Galileo, the first astronomer to observe the planet through a telescope.

Unsurprisingly, it has the least amount of water of any solar system object.

More distant Europa doesn't have that problem. Also tidally heated, but nowhere near as intensely, ice is turned to water. A lot of water. There is thought to be more liquid water on Europa than in all of Earth's oceans, lakes, rivers and seas put together. That puts it squarely towards the top of the list of places to go searching for life elsewhere around the sun.

Saturn

The last of the classical planets known to our ancient ancestors, Saturn sits almost 1.5 billion kilometres from the sun. The fact that we can see it with our own eyes from such a distance is testament to its size and how much sunlight it reflects back our way. The second-biggest planet, you could fit over 750 Earths inside.

But it is incredibly light for its size. An average density of 0.7 grams per cubic centimetre is the lowest of any planet. That's less dense than water (1 g/cm^3), meaning Saturn would float in a bathtub if you fashioned one big enough. Although, in reality, it would freeze the water – the average temperature on Saturn is -178 degrees Celsius.

The planet's distinctive yellow colour comes from crystals of ammonia high up in its atmosphere. Storms are occasionally seen roaring through that atmosphere and are more prevalent when Saturn reaches its closest point to the sun every thirty years or so.

With the arrival of the *Voyager* probes, astronomers noticed a hexagonal cloud pattern atop the planet's north pole. Each of the hexagon's six sides is greater than the diameter of Earth. The *Cassini* spacecraft observed it too, and between 2013 and 2017 it changed colour from blue to gold.

As with Jupiter, astronomers are unsure what is going on beneath Saturn's cloud decks. Clouds of water are thought to exist beneath the ammonia crystals. Beneath that there may be a layer of metallic hydrogen, followed by a dense, rocky core between nine and twenty-two times the mass of Earth. Some scientists have even speculated that diamonds form in Saturn's atmosphere at the rate of one thousand tonnes a year. Lightning turns methane gas into carbon dust, which then gets crushed into diamonds as it falls towards the planet's core.

Rings

Saturn's enigmatic rings are far and away the most famous part of the solar system. Yet, despite much scrutiny, no one knows exactly where they came from.

While they may look solid from a distance, they are made up of individual chunks of ice that can be as big as houses. If you were to collect all of the ring material together in one ball, you'd end up with something similar in size to Saturn's moon Mimas. So perhaps they started life as a satellite that was ripped apart by the planet's gravity or smashed to bits in a collision.

Recent data from the Cassini mission indicate that the rings must be very young compared to the age of the solar system, maybe just 100 million years old. If they were older, the solar wind would have darkened the ring material significantly. We are very lucky to live at a time when Saturn has rings, because it seems that it hasn't for the vast majority of its history.

Even with an amateur telescope from Earth you'd notice the rings have gaps. The largest is called the Cassini division and Mimas also features here – the moon's gravity keeps the gap open. Some of Saturn's moons orbit inside the rings – those that do are called 'shepherd moons'. The rings are labelled with capital letters, but in alphabetical order of their discovery, not their distance from the planet.

NASA's *Cassini* probe captured this stunning image of the sun lighting Saturn's rings from behind.

The rings still hold many mysteries. Since the *Voyager* fly-bys in the early 1980s, astronomers have noticed dark blotches in the rings – shadowy features spreading out like spokes on a bicycle wheel. They've been photographed again with the more recent *Cassini* mission, but we still don't know what they are.

CASSINI

Cassini revolutionized our understanding of the ringed planet. Launched in 1997, it arrived at Saturn in 2004. On 15 September 2006, it took one of the most spectacular images in the history of astronomy. It shows Saturn eclipsing the sun, with sunlight backlighting the planet's jaw-dropping ring system (see page 105). As if that weren't enough, a tiny dot – which you might mistake for one of Saturn's moons – is actually the Earth over a billion kilometres distant.

In 2017, running low on fuel, *Cassini* undertook a series of more than twenty daring swoops down through the rings as the mission drew to a close. It got closer to the rings than ever before at speeds of 100,000 kilometres per hour. Astronomers even had time to snap another picture of the Earth in the distance through the ring material.

In one last hurrah, astronomers deliberately crashed *Cassini* into Saturn in September 2017, twenty years after it departed Earth for the outer solar system. This prevented it from inadvertently contaminating Saturn's rings or moons.

Moons

Like Jupiter, Saturn is escorted around the sun by over sixty natural satellites. Most are tiny, but Titan is bigger than Mercury. It's the second largest moon in the solar system after Jupiter's Ganymede.

Mimas is nicknamed 'The Death Star' for its resemblance to the moon-sized space station in the *Star Wars* franchise. This eerie match is pure coincidence – the Death Star appeared on screen three years before the first images of Mimas.

Hyperion is Saturn's oddest-looking moon. A giant cosmic pumice stone, ridden with holes and irregular in shape, it was the first non-round moon to be discovered. Perhaps it's a fragment of an ancient collision.

But the moon that has everyone's attention right now is Enceladus. Liquid water is gushing out of cracks in the surface ice. In 2017, astronomers announced they'd also found complex chemistry there – the building blocks of life. Water plus building blocks might equal life itself, although the fervour was dampened when toxic methanol was also detected. Nevertheless, Enceladus sits with Europa at the top of the hotlist for potentially habitable moons.

That leaves Titan. Along with its size, it stands out because it boasts a thick, hazy atmosphere – the only moon in the solar system to do so. Astronomers despatched the *Huygens* lander down through the murk to the surface. It landed in a dried-up riverbed on 14 January 2005 in a region called Xanadu. It's the only thing we've ever landed in the outer solar system.

Surface maps of Titan reveal a world that looks all too familiar. Coastlines, archipelagos, islands and peninsulas have been chiselled away by oceans lapping at ancient shores. Except it's far too cold this far from the sun for that eroding liquid to be water. The culprit here is liquid methane.

Uranus

We think people living near our poles have it bad. For large parts of the year they are plunged into permanent darkness or experience days that never end (see page 72). However, the illumination of Uranus's poles takes things to a whole new level.

The planet is tipped over on its side, its poles roughly in line with the plane of its eighty-four-year orbit around the sun. That means the poles of Uranus experience forty-two years of constant day followed by forty-two years of relentless night. Not that the days are particularly bright. At twenty times further from the sun than us, the intensity of sunlight there is just one four-hundredth of what we receive. The planet is four times the width of Earth.

No one knows exactly how Uranus ended up so topsy-turvy but, as with most oddities in the solar system, the finger is pointed at a giant impact. Uranus's ring system also follows this tilt, meaning the rings appear to straddle the planet almost top to bottom rather than side to side like Saturn's.

Both Uranus and Neptune are sometimes referred to as 'Ice Giants', their chemical composition marking them out as different from the gas giants Jupiter and Saturn. Water, ammonia and methane all turn to ice this far from the sun.

So far twenty-seven moons have been discovered in orbit around Uranus. All are named after characters from Shakespeare plays or works by Alexander Pope. Familiar names include: Romeo, Juliet, Ophelia (*Hamlet*), Puck and Oberon (*A Midsummer Night's Dream*).

Titania (*A Midsummer Night's Dream*) is the biggest, but at just shy of 800 kilometres wide it's less than half the diameter of our moon. Miranda (*The Tempest*) is perhaps Uranus's most distinctive satellite, sporting huge scars across its surface. These suggest it is somewhat of a 'Humpty-Dumpty moon' – smashed apart and not quite put back together again.

Voyager 2 is the only mission to have visited Uranus. Sidling up to the planet in 1986, it observed an almost featureless blue-green atmosphere in stark contrast to its more active neighbours. There have been calls to send a new mission to explore this overlooked world in an attempt to decipher some of its secrets.

Neptune

During the summer of 1989, *Voyager 2* left the giant planets behind. Heading away from the sun, scientists turned its cameras back for a last, fleeting glimpse of Neptune and its largest moon, Triton. Both were illuminated as beautifully thin crescents as the last glimmer of reflected sunlight disappeared from view.

By contrast to Uranus, Neptune's sea-blue surface is incredibly dynamic. *Voyager 2* spotted a Great Dark Spot the width of the Earth in the planet's southern hemisphere. It was accompanied by a fast-moving bright feature affectionately known as Scooter. By the time the Hubble Space Telescope took another look at Neptune in 1994, the Great Dark Spot had gone, replaced by a new storm north of the equator. Neptunian storms rage with the strongest winds in the solar system – up to 580 metres per second (1,300 miles per hour).

In terms of mass, Neptune sits roughly between the Earth and Jupiter – seventeen times heavier than the former and nineteen times lighter than the latter. It orbits thirty times further from the sun than us and takes 165 years to complete one orbit. Temperatures drop as low as –218 degrees Celsius.

So far we've discovered fourteen Neptunian moons, the last as recently as 2013. By far the most interesting – and perplexing – is Triton. Not only is it geologically active – putting it in a rare club alongside Jupiter's Io and Saturn's Enceladus – it also orbits Neptune in the opposite direction to the planet's motion around the sun. It's the only large moon in the solar system with this so-called *retrograde* orbit.

That makes explaining where it came from extremely difficult. Normally, retrograde moons are small – objects such as asteroids or comets that came close to a planet at a jaunty angle and ended up circling it backwards. It's not so easy to get a 2,700-kilometre-wide object to do the same. Many astronomers believe Triton must have been a dwarf planet (see page 112) drawn in from further out by Neptune's gravity. Why it didn't crash into Neptune, instead of neatly settling into a stable orbit, is unknown.

Much of Triton's western half has a strange-looking surface that astronomers refer to as 'cantaloupe terrain' due to its resemblance to a melon. Its south pole is capped with nitrogen and methane ice that has been peppered with dust deposits ejected from cryovolcanic geysers.

Pluto

Poor Pluto. This frozen snowball beyond the orbit of Neptune has been through the wringer. When it was discovered in 1930, by American astronomer Clyde Tombaugh, it was immediately lauded as the ninth planet. But its membership of that exclusive club lasted less than a century.

It all started to go wrong in the mid-2000s. First, astronomers found Eris, a more distant world thought to be bigger than Pluto and also orbiting the sun. Astronomers faced a dilemma over how to classify Eris. If Pluto is a planet then Eris has every right to be, too – it also orbits the sun and is larger. Where is the line between the planets and the rest of the solar system's smaller objects?

Pluto's status faced further scrutiny as it stands out for other reasons. For starters, it crosses orbits with Neptune. For twenty years of its 248-year orbit it is closer to the sun. Between 1979 and 1999 it was the eighth, not ninth, planet. It is also locked into a gravitational dance with Neptune that astronomers call *resonance*. Pluto makes exactly two orbits of the sun for every three Neptune makes. This means they're always kept apart and there's no risk of a collision.

Pluto has an equally dysfunctional relationship with its largest moon – Charon. Unlike a normal planet–moon system, these two worlds orbit a common point in the empty space between them.

So something had to give. In the summer of 2006, the decision was made at a meeting of the International Astronomical Union (IAU) to reclassify Pluto (and Eris) into a new class of object – the dwarf planets (see next section).

Pluto falls down because it hasn't 'cleared the neighbourhood around its orbit' – it's not the biggest thing in its path around the sun (Neptune is). The decision remains highly controversial.

Back in 2005, when it was still a planet, NASA launched the *New Horizons* probe to explore Pluto. By the time it got there in 2015, its destination was a planet no longer. Planet or not, the probe revealed a fascinating place far beyond the expectations of many astronomers. Not only did we get the first high-resolution images of this frigid world, it is also more active than anyone suspected. Despite temperatures there dipping as low as -240 degrees Celsius, some unknown geological process has reshaped its surface in the recent past.

In preparation for the arrival of *New Horizons*, astronomers scoured the area around Pluto for any sign of additional moons that could be hazardous to the mission. They found two – Kerberos and Styx – joining the three established moons of Charon, Nix and Hydra.

The dwarf planets

The decision to introduce a strict definition of planet-hood, along with the rules for what makes a dwarf planet, tore up the textbooks. Here's part of the resolution passed by astronomers at a meeting of the IAU in Prague in August 2006:

(1) A planet is a celestial body that:
    (a) is in orbit around the sun,
    (b) has sufficient mass for its self-gravity to overcome

rigid body forces so that it assumes a hydrostatic equilibrium (nearly round) shape, and

(c) has cleared the neighbourhood around its orbit.

(2) A 'dwarf planet' is a celestial body that:
(a) is in orbit around the sun,
(b) has sufficient mass for its self-gravity to overcome rigid body forces so that it assumes a hydrostatic equilibrium (nearly round) shape,
(c) has not cleared the neighbourhood around its orbit, and
(d) is not a satellite.

Pluto falls down on point 1(c). Its loss was Ceres's gain. The largest asteroid in the main belt, it was also considered a planet back when it was discovered in 1801. It lost that honour pretty soon afterwards, but the 2006 resolution saw it promoted to a dwarf planet. Three worlds more distant than Pluto – Eris, Haumea and Makemake – were also anointed as dwarf planets.

Eris is the largest solar system body not visited by a spacecraft. It sits nearly 100 times further out than Earth on an orbit that takes 558 years to complete. It is accompanied by at least one moon – Dysnomia.

Haumea ('HOW-MAY-A') is an odd, egg-shaped ovoid tumbling around the sun with satellites Hi'aka and Namaka for company. In 2017 astronomers discovered it has a ring. It's named after a Hawaiian goddess as it was discovered using telescopes on the island. However, it was originally nicknamed 'Santa' because it was discovered just after Christmas.

Makemake ('MAH-KAY-MAH-KAY') derives its name from Easter Island mythology – it was discovered at Easter. Hence its original nickname: 'Easter bunny'. A moon was announced in 2016, but at the time of writing was yet to be officially named.

In reality there are far more than five dwarf planets. Sedna, for example, is almost certainly one. However, it's on an 11,400-year orbit. That makes it hard to test accurately if it meets the criterion of being nearly round in shape. So expect more dwarf planets as we build bigger telescopes and get a closer look at more of these tiny, far-flung worlds.

The Kuiper Belt and Scattered Disc

The discovery of Pluto in 1930 fired imaginations into overdrive. Astronomers began speculating that the new planet was just one of a whole host of other worlds orbiting the sun beyond Neptune. Over the decades, various different thinkers riffed on variations of this idea, but Dutch astronomer Gerard Kuiper's name has become most associated with the region. Today we know it as the Kuiper Belt.

That seems a bit harsh, particularly because Kuiper explicitly said the belt no longer existed. Irish astronomer Kenneth Edgeworth was closer to the truth and published his ideas before Kuiper. Nevertheless, when the first object since Pluto was discovered beyond Neptune in 1992, it was lauded as proof of a Kuiper Belt, not an Edgeworth Belt.

The Kuiper Belt stretches from the orbit of Neptune out to around fifty-five times further from the sun than the

Earth. Over a thousand Kuiper Belt Objects (KBOs) have been found so far and astronomers believe there could be as many as 100,000 of them over 100 kilometres in diameter. However, their total mass is no more than a tenth of the Earth. They're believed to have formed from planetesimals in the same way as planets, but are much smaller because there were fewer building blocks that far from the sun. Pluto is the most famous KBO, along with the dwarf planets Haumea and Makemake.

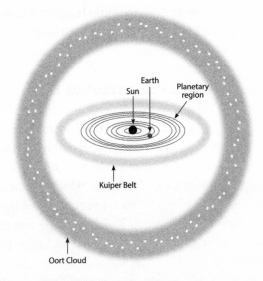

The solar system does not end with the planets. Further out than Neptune sit the Kuiper Belt and Oort Cloud.

Eris, the other trans-Neptunian dwarf planet, lies beyond the Kuiper Belt in a region known as the Scattered Disc. The orbits of objects in this region can carry them one hundred times further from the sun than the Earth. The origin of the Scattered Disc hasn't been conclusively nailed down, but most astronomers point the finger at Neptune – the planet scattering objects from the Kuiper Belt as it moved away from the sun early in the solar system's history (see page 121).

Planets Nine and Ten

'My Very Easy Method Just Speeds Up Naming Planets' – it used to be so simple to remember the order of the solar system's nine planets. Then Pluto was demoted and we had to come up with alternative mnemonics. Well, be prepared to change them yet again.

The more we've explored the trans-Neptunian objects, the more we've sensed something rather strange going on. It was noted in 2014 that two Kuiper Belt Objects – Sedna and 2012 VP$_{113}$ – shared very similar orbits. Specifically, they had a common *argument of perihelion* – the angle

The similarly angled orbits of several small objects in the outer solar system could be the result of the gravity of an unseen ninth planet.

at which they're inclined to the planets at their closest approach to the sun. This value should be reasonably random, but they are eerily twinned. Then, in 2016, astronomers revealed that four additional objects share the same feature. The likelihood of that being down to random chance was calculated at just 0.007 per cent.

The leading explanation is that we've missed a planet in our solar system. Much as we discovered Neptune based on what its gravity is doing to Uranus (see page 45), the pull of this as-yet-unseen planet might be lining up the orbits of these six small worlds. The planet would have to be ten times the mass of Earth and take 10,000 to 20,000 years to orbit the sun. Astronomers are currently frantically scanning the skies on the hunt for it, the first new planet in almost two centuries.

There may even be two. In June 2017, astronomers published research suggesting that a tenth planet (if Planet Nine is indeed there) might explain the warped orbits of some other Kuiper Belt Objects. Much smaller than Planet Nine, it would need to be about the mass of Mars.

It is clear that our picture of the solar system is far from complete and could easily change again in the years ahead.

The Voyagers and the heliosphere

Just where does the solar system end? One way to define an edge is to talk about where the magnetic influence of the sun begins to wane. And thanks to the *Voyager* probes we have accurate data on the boundary of this so-called heliosphere.

While *Voyager 2* was busy exploring Uranus and Neptune, *Voyager 1* was ploughing ahead towards the fringes of the solar system. Today it is more than 20 billion kilometres from the sun. *Voyager 2* is following on about 3 billion kilometres behind. While there is nothing to see out there in the inky void of space, the scientific instruments of both probes are still active. They've been sending home daily measurements of the solar wind at their backs. It takes more than thirty hours for those radio signals to return to Earth.

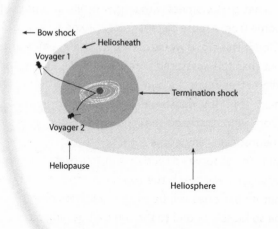

In 2012 the *Voyager 1* probe became the first human-made object to leave the solar system when it crossed the heliopause.

Remarkably, from 2010 onwards, the speed of the solar wind felt by *Voyager 1* dropped to zero. By August 2012, astronomers were confident enough to declare that the probe had crossed over the heliopause – the edge of the heliosphere – and moved into interstellar space. It was celebrated as the first emissary of Earth to leave the solar system. At its current speed it would take around 30,000 years to reach the next solar system.

THE OORT CLOUD

Short-period comets – with orbits under 200 years – originate from the Kuiper Belt and Scattered Disc (see page 114). Examples include Halley's Comet and Comet 67P. Comets with longer orbits are thought to come from a more distant reservoir known as the Oort Cloud. This cloud would start a thousand times further out than the Scattered Disc and continue to 30 *trillion* kilometres from the sun – more than halfway to the next star. It's named after Dutch astronomer Jan Oort, who discussed the idea in the 1950s. It remains theoretical – any comets there are too far from the sun to observe directly with current telescopes. *Voyager 1* will reach the region in around 300 years, but its batteries will be long dead. The comets would be so loosely bound to the sun that passing stars could give some a nudge, sending them inwards towards us. That sees them return from where they came – astronomers believe they formed in the inner solar system before being scattered far and wide by the movements of the giant planets to their current orbits.

Indeed, the achievements of *Voyager 1* should go down in history as a valuable first. But claims it has left the solar system should be taken with a word of caution. It has only left the solar system if you define its edge in terms of magnetism. It is likely there are significant objects further out which still orbit the sun, not least the potential Planet Nine. Can *Voyager 1* really have left the solar system if it hasn't yet reached the distance of the sun's outermost planet?

The Nice model

The modern solar system is a complicated place with many complex features. Piecing together a coherent picture of how you get from a disc of debris surrounding the infant sun to an intricate system of planets, moons, dwarf planets, asteroids and comets is no mean feat. Efforts have been greatly bolstered in recent years by the advent of supercomputers capable of churning through highly detailed models. The best fit to emerge so far is known as the Nice model, named after the city in France where it was put together.

The model suggests that the four giant planets started out in a more tightly knit group, before gravitational interactions saw them migrate to their current positions. You get the best match to today's solar system if Jupiter moved inwards and the other three gas planets headed outwards. In some models Uranus and Neptune even swap order. As Jupiter encroached on the asteroid belt it would have scattered many space rocks, perhaps accounting for the Late Heavy Bombardment (see page 86). Similarly,

Neptune pushing outwards disrupted the Kuiper Belt to form the Scattered Disc. It could also have drawn its large, irregular moon, Triton, in from the belt (see page 110). Some comets were scattered far enough to form the distant Oort Cloud.

Originally, the Nice model started out with four giant planets. But astronomers also ran the model with a fifth giant planet of various sizes. To their surprise, it produced a solar system that looked even more like ours. If there really was a fifth giant planet, where is it now? If it was also scattered by its bigger neighbours then it could have ended up marooned far out beyond Neptune. If Planet Nine is indeed found, it is almost certainly the missing planet (see page 116).

The only problem with a Nice model with five giant planets is that recent simulations have shown their migrations would have had a disastrous effect on the rocky planets. In almost all models Mercury ends up ejected from the solar system entirely. That clearly hasn't happened. You can save the five-giant-planet Nice model by saying the migration of those planets must have occurred before the rocky planets formed. But then you can't explain the Late Heavy Bombardment as the result of Jupiter's migration. That would suit those who argue the LHB never happened (see page 86).

So our ideas about how our solar system formed are constantly evolving as we piece together the latest computer simulations, with new discoveries in the expanse beyond Neptune.

CHAPTER 4

Stars

How bright?

Just a quick glance at the night sky will reveal that some stars are brighter than others. Astronomers have a measure of how bright a star appears to us – *apparent magnitude*. The system is based around the star Vega, one of the brightest stars in the night sky. It is said to have an apparent magnitude of zero. Any star with a negative apparent magnitude is brighter than Vega and those with a positive value are dimmer. Each step on the scale is approximately equal to a 2.5 times brightness difference. So a -1.0 magnitude star is 2.5 times brighter than Vega, a +2.0 star is 6.25 times dimmer (2.5 × 2.5).

However, the brightest objects in the night sky are not stars: the full moon (-12.74), International Space Station (-5.9), Venus (-4.89), Jupiter (-2.94) and Mars (-2.91) are all more dazzling. Sirius is the brightest star at -1.47.

Just because a star appears bright in our night sky, it isn't necessarily particularly bright in reality. It could just be very close to us. Similarly, a staggeringly bright star might appear dim to us because it is really far away. So astronomers have an alternative measure for the true brightness of stars called *absolute magnitude*. It tells you

VARIABLE STARS

Not all stars have a fixed brightness – their apparent magnitude appears to change over time. Astronomers call them variable stars. Stars normally vary for one of two reasons: either they really change their brightness, or there's something periodically getting in the way.

One of the most famous variable stars is Algol – also known as the Demon Star. On star maps it is often drawn as the evil eye on the severed head of Medusa held aloft by the hero Perseus. Every 2.86 days it drops from magnitude 2.1 to 3.4 for a period of about ten hours. That's because it isn't a single star, but a system of three. It appears to dim when one of the fainter stars partially eclipses the brightest.

Cepheid variables are another famous type of varying star. Polaris – the Pole Star, or North Star – is the nearest example to the Earth. These stars periodically expand and contract, causing them to get brighter and dimmer in a repeating pattern.

how bright the star would appear if you were a distance of 32.6 light years away from it. It's measured on the same scale as apparent magnitude.

Sirius is a classic example. Its apparent magnitude may be a bright -1.47, but its absolute magnitude is 1.42. It appears to be the brightest star in the night sky only because it is relatively close to us. Rigel, a star in the adjacent constellation of Orion, has an apparent magnitude of 0.12 but an absolute magnitude of -7.84. It is one of the most inherently luminous stars you can see in the night sky.

How far?

Working out the absolute magnitude of a star requires us to know how far away it is. That way we can work backwards from its apparent magnitude. But there is no way to take a tape measure to the heavens, so how are such distances calculated? For the nearest stars to us, including many of those visible in the night sky, astronomers use a technique called *parallax*.

To see how this works we're going to replace the star in question with your index finger. Raise your index finger at arm's length, close one eye and line your finger up with an object in the distance – perhaps the edge of a picture frame or the corner of a room. Now change the eye that's open. You should see your finger jump to one side. Next, repeat the whole exercise but with your finger much closer to your face. Has it jumped more or less than before?

Hopefully you'll see your finger has jumped considerably more. When a nearby object is seen from two vantage points (your two eyes in this case) it will jump more against the background than a more distant object. Astronomers replicate your two eyes by observing a star at intervals six months apart, when Earth is on opposite sides of the sun. A nearby star jumps a lot more against the background than one that's further away. Trigonometry is used to convert the angle through which the star jumps into a distance to it. The European Space Agency's Gaia telescope, launched in 2013, can use parallax to measure the distances to stars within a few tens of thousands of light years of Earth. Further than that and the angle becomes too small to measure accurately and astronomers use another method to measure cosmic distances (see page 182).

How hot?

Your bathroom taps have been lying to you your entire life. Every day we wash our hands and brush our teeth confronted by a sink that insists red is hot and blue is cold. Really it is the other way around. And you don't need to look at the stars to see it. The hottest flames, such as those produced by a blowtorch, are blue. A normal, open flame is yellow. Only when a fire starts to cool down and die out does it then glow red.

Stars are not on fire, but the principle is the same. By looking at a star's colour we can tell how hot it is. The coolest stars are red, with surface temperatures around 3,000K (Kelvin; to convert to degrees Celsius subtract 273). Yellow stars sit somewhere in the middle with surface temperatures of 6,000K. The hottest stars, which appear blue, can reach temperatures of 50,000K.

Astronomers split stars into seven groups based on their colour using a system known as the Harvard spectral classification. The groups are given the letters O, B, A, F, G, K and M. Originally the groups ran from A to Q but it turned out there were significant overlaps between them and many of the groups were dropped.

Spectral class	Colour	Temperature (K)	Per cent of all stars
O	Blue	>30,000	0.00003
B	Blue-White	10,000–30,000	0.1
A	White	7,500–10,000	0.5
F	Yellow-White	6,000–7,500	3
G	Yellow	5,200–6,000	7.5
K	Orange	3,700–5,200	12
M	Red	2,400–3,700	76.5

The sun is a G-class star, so the majority of the stars in the universe are cooler than ours. The brightest O-class star in the night sky is Alnitak in Orion's Belt. M-class stars are too dim for your eyes to see.

The percentages shown here refer to stars in the main part of their lives – astronomers would say they are *main sequence* stars as they fall on the diagonal line of a Hertzsprung–Russell diagram.

The Hertzsprung–Russell diagram

The H–R diagram is the iconic graph in astronomy. It plots the absolute magnitude of stars against their colour (or spectral class). It was created in the early twentieth century by Danish astronomer Ejnar Hertzsprung and American astronomer Henry Norris Russell to visualize the evolution of stars.

Small, cool stars (K and M class) are found in the bottom right-hand corner of the graph. Bigger, hotter stars (O and B class) are located towards the top left. The line between these two extremes is known as the main sequence. Stars on this line are fusing hydrogen into helium in just the same way as the sun (see page 51).

However, a star runs out of hydrogen in the core as it ages. We'll see what happens in more detail later, but the star swells up. In doing so it spreads its heat over a much greater surface area and so its colour reddens. Astronomers say it has 'evolved off the main sequence' and these red giants and red supergiants are found above the line.

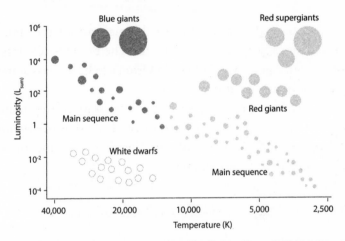

The Hertzsprung–Russell diagram shows the relationship between a star's temperature and its luminosity. Stars spend most of their lives on the 'main sequence'.

How big?

Stars come in different sizes and masses, and astronomers have discovered a strict relationship between the mass of a star and its luminosity. This is called the *mass–luminosity relation* (see graph on next page). The more massive the star, the greater its inherent brightness (absolute magnitude).

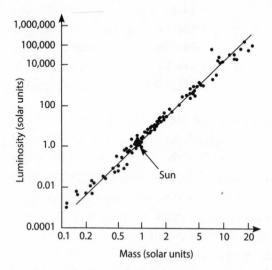

Astronomers have noticed a strict relationship between a star's mass and its brightness (luminosity). This allows them to weigh new stars through their brightness.

To work out the mass of a newly discovered star, astronomers measure its apparent magnitude and work backwards using its distance to calculate its luminosity (absolute magnitude). The graph of the mass–luminosity relation then gives its mass (see table opposite). High-mass stars are found near the top left on the main sequence on the H-R diagram, low-mass stars to the bottom right. R136a1, a star in the Large Magellanic Cloud, is the most massive and most luminous known. It is 315 times heavier than the sun.

Spectral Class	Mass (in multiples of the sun's mass)
O	> 16
B	2.1–16
A	1.4–2.1
F	1.04–1.4
G	0.8–1.04
K	0.45–0.8
M	0.08–0.45

Astronomers can also calculate the size of stars using Stefan's law – named after physicist Josef Stefan (1835–93). He worked out that the amount of energy a hot object radiates per second depends on its size and its temperature. When it comes to a star, we know how much energy it radiates per second – that's its luminosity. We also know its temperature from its colour. So we can use Stefan's law to calculate the star's size.

The largest-known star is UY Scuti. Found in the small constellation of Scutum, the Shield, it has an estimated width of 1,708 suns. That means roughly 5 billion suns would fit inside its volume. If it replaced the sun in our solar system, its surface would sit between the orbits of Jupiter and Saturn.

How old?

In the early universe, before any stars formed, the only elements around were hydrogen and helium. Then the first stars ignited and began to fuse some of the hydrogen into more helium in the same way the sun does now (see page 51). When those stars grew old and evolved off the main sequence they began to churn the helium into even heavier elements such as carbon, nitrogen, oxygen, silicon and iron (see page 135). These massive stars exploded at the end of their lives as brilliant supernovas, flinging those heavier elements far out into the universe. Some of those elements ended up being incorporated inside new stars.

So astronomers can age stars by looking at their chemical composition. The oldest stars are made only of hydrogen

OPEN AND GLOBULAR CLUSTERS

On a clear night, far from city lights, you'll be dazzled by 3,000 stars. Most appear to exist in isolation, but you'll notice a few grouped together. Cast a pair of binoculars across the sky and you'll see even more of these star clusters, particularly along the band of the Milky Way.

Astronomers divide them into *open* clusters and *globular* clusters. Open clusters are quite loose-knit, whereas globular clusters appear more like a blob. However, their biggest difference lies in the ages of their members. Open clusters tend to contain very young stars, whereas the stars in globular clusters are ancient.

Take the most famous open cluster: the Pleiades in Taurus (also known as the Seven Sisters). Its stars are only 100 million years old. Compare that with the stars in M13 (the Great Globular Cluster in Hercules) which are over 11 billion years old. Replace the age of the universe with the average life expectancy of a human and the M13 stars would be approaching retirement while the Pleiades would still be in nappies.

and helium, formed at a time when that's all there was. The youngest stars were formed at a time when there was a much wider array of ingredients on offer – they're more chemically diverse. Astronomers have a measure for this, called metallicity. Unlike chemists, astronomers consider any element other than hydrogen and helium to be a metal. A low metallicity means the star is old and pristine. The higher the metallicity, the younger the star. The sun's metallicity is 0.02: 2 per cent of the sun's mass comes from elements other than hydrogen and helium.

Of course, this technique relies on knowing what a star is made of. For that, astronomers use spectroscopy. If you take the light from a star and pass it through an instrument called a spectrometer (which is a bit like a prism) you end up with similar black lines to those seen by Fraunhofer in the spectrum of the sun (see page 50).

These are absorption lines – missing colours – caused by different elements within the star swallowing that particular colour of light. That hue never made it out to travel to the Earth. This spectrum looks a bit like a colourful barcode and it does exactly the same job – it carries information about what's inside the star and, in turn, how old it is.

The life cycle of stars

Star birth

Just like people, stars are born, grow old and die. Stars form from vast, beautiful pillars of gas called molecular clouds that are incredibly sparse. Place a little cube just a centimetre on each side into a molecular cloud and it would have around one hundred gas molecules inside. The same cube on Earth contains 100,000 trillion air molecules. Put the cube in the heart of star and it would have a 100 trillion trillion particles in it.

So how do you get from something as loosely bound as a molecular cloud to something compact enough to fuse hydrogen into helium (the hallmark of a star)? You tip the balance in favour of gravity. British astronomer James Jeans (1877–1946) calculated the maximum mass a molecular cloud can have before gravity takes over and it begins contracting. Astronomers call this the *Jeans mass*. It also depends on the temperature and density of the cloud.

The contraction can be triggered by external events. Maybe two molecular clouds merge and their combined mass surges above the Jeans mass. Or perhaps a star explodes nearby, sending an intense shock wave through the cloud, marshalling the gas closer together until gravity does the rest.

As a molecular cloud contracts, it splinters into smaller sections. These shrinking sections – called protostars – begin to spin faster and faster, much as when an ice skater draws in their arms. The temperature and pressure continue to increase until hydrogen fuses into helium inside a rotating sphere of a gas: a star is born. The whole process takes tens of millions of years.

Astronomers can observe star formation happening in regions such as the Orion Nebula – a bright star factory visible with the unaided eye as a fuzzy smudge underneath the three stars of Orion's Belt. It is the nearest stellar nursery to the Earth. Dark, flat discs of debris have also been observed around some of these infant stars. They're called proto-planetary discs (or proplyds) and it's thought gravity will sculpt these discs into planetesimals, then planets.

Red giants

A star uses up more and more hydrogen as it ages, until eventually its fusion rate starts to drop. This means there isn't as much energy being produced to prop up the core against gravity. The core contracts, the temperature rises and the fusion rate increases. This *main sequence brightening* has seen the sun get around 30 per cent brighter since it formed 4.6 billion years ago.

It will continue getting brighter and hotter, until in a billion years the temperature on Earth will rise well beyond 100 degrees Celsius. Our planet will become a scorched and sterilized rock, its oceans boiled away. The sun – the giver of life – will ultimately be its exterminator.

In 5 billion years, hydrogen fusion in the core will cease entirely and it will shrink, rocketing the temperature from 15 million degrees to around 100 million degrees. Hydrogen fusion will restart in a shell surrounding the super-hot core. The restarting of fusion marks the point where the sun begins to move off the main sequence of the Hertzsprung-Russell diagram – the graph that tracks the evolution of stars (see page 126).

The energy from this reinvigorated fusion will see the outer layers of the sun bloat outwards until it becomes a hundred times wider than it is today. Mercury will succumb to its fiery embrace. Venus might, too. With its heat spread over a much greater surface area, our sun will redden. It will have become a *red giant*. At over two thousand times more luminous than it is today it will easily melt metal on the surface of the Earth. Our planet might even be dragged into the sun's outer layers.

Planetary nebulae and white dwarfs

In the core of red giants, the increased temperature sees helium fused into carbon and oxygen. But for small stars – those less than eight times the mass of the sun – the temperature and pressure isn't sufficient to fuse carbon. Once all the helium is exhausted, a dense carbon-oxygen core about the same size as the Earth remains. Astronomers call this object a *white dwarf*. With no new source of heat, it eventually cools and fades away to become a black dwarf.

By this point the outer layers of the red giant have been blown out into space by strong stellar winds. This is not an explosion – it isn't as powerful as that. The gas forms shells around the central white dwarf. Astronomers call these objects *planetary nebulae*. However, they have nothing to do with planets. When astronomers first saw them through early telescopes the gas shells made them appear similar in shape to planets. Our understanding of them has changed, but the unhelpful name has persisted.

Planetary nebulae are some of the most visually stunning objects in the night sky. Famous examples include the rainbow-coloured Ring Nebula in the constellation of Lyra and the Cat's Eye Nebula in Draco. Look at photos of these glorious gas clouds and you'll be able to spot the white dwarf hiding in the centre.

Red supergiants

Stars with a mass greater than around eight to ten suns evolve in a different way. To start with, the process is similar, but then it diverges dramatically. Initially they bloat into even bigger stars than red giants. These *red supergiants* can swell to over a thousand suns wide. They are also much brighter than red giants. Some of the brightest stars in our night sky, including Betelgeuse in Orion and Antares in Scorpius, are in this phase of their lives. Replace the sun with Antares and its outer edge would sit beyond the orbit of Mars. Other red supergiants would reach Jupiter and even Saturn.

It's in the core where things are very different. The sheer size of these stars means the core temperature rises to the point where carbon fusion kicks in, creating magnesium and oxygen. When the core runs out of carbon it contracts further, the temperature rises again and oxygen starts to fuse into silicon and neon. And so it keeps going – each time an element is exhausted the core shrinks and the temperature rises, allowing a new element to fuse. The pace quickens, with each stage more short-lived than the last. A massive star might have been burning hydrogen into helium for 10 million years, but the final phase of fusing silicon into iron lasts a single day.

But there the process has to stop. Iron is the most stable element in the periodic table and it cannot be fused. The core ends up looking like an onion, with a mass of iron in the centre surrounded by concentric layers of other unspent elements. Now, with nothing to support the star against gravitational collapse, its fate is sealed.

Supernovae

In the year 1054, Chinese astronomers wrote about the unexpected arrival of what they called a 'guest star'. Appearing as if from nowhere, it was so bright that it could be seen during the day for almost a month. It gradually faded from the night sky, disappearing completely after nearly two years.

We now know they had witnessed a supernova explosion – one of the most violent and energetic events in the universe. Modern astronomers have identified the remnants of this cataclysm as the Crab Nebula in the constellation of Taurus. Gas is still hurtling away from the explosion at 1,500 kilometres per second nearly a thousand years later. The death throes of a massive star, a

The famous Crab Nebula (M1) in the constellation Taurus. It is the remnant of a supernova explosion that detonated in 1054.

supernova can shine as bright as 10 billion suns and release more energy than the star did during its entire life.

It starts with the dense iron core that forms at the heart of a red supergiant. Unable to resist gravity, the core collapses in on itself in less than a second at almost a quarter of the speed of light. This sends a shock wave outwards almost as quickly, tearing through the star's outer layers and blasting it apart.

The force of the explosion slams atoms into their neighbours, building up the elements heavier than iron. A supernova rockets both the elements forged by fusion and the explosion itself into interstellar space. This enriches molecular clouds with a wide variety of elements, which then become part of any stars and planets later made there.

Are you wearing any jewellery? The gold, silver and platinum were all fashioned in a supernova (and in neutron-star collisions). The iron in your blood and the oxygen it helps carry around your body were made inside massive stars by fusion and then blasted across the universe by a supernova. They are the ultimate cosmic recyclers and without them we wouldn't be here.

Neutron stars and pulsars

Deep inside the Crab Nebula sits the smoked-out ruins of a once mighty star. After its dense iron core had buckled under its own weight, it almost collapsed to nothing. The iron was broken down under great pressure and eventually it all turned into neutrons – the neutral particles found at the centre of atoms. Stars between eight and thirty times more massive than the sun end up this way, surrounded by a supernova remnant.

However, there is a limit to how tightly neutrons can be packed together. This led to the collapse levelling out when the core had shrunk to a super-dense mass just 30 kilometres across. All that remains of a colossal red supergiant star once 100,000 times wider than Earth is a ball smaller than London. Huge amounts of mass are squeezed into such a small space that a single teaspoon of material from a neutron star weighs 10 million tonnes.

As the star condensed, its spin ramped up. Once, it may have rotated every few weeks. Now it spins thirty times a second. The star's magnetic field gets more concentrated, too, becoming a trillion times more powerful than Earth's. That marshals super-heated material into powerful beams that rage away from the neutron star's poles.

This makes a neutron star the cosmos' equivalent of a lighthouse. If we happen to be in the direction of the rotating beams, we pick up regular, repeating bursts of radio waves. So we call these objects pulsars – from the contraction of 'pulsating star'.

Pulsars keep time so well that when the first one was discovered in 1967, by Anthony Hewish and Jocelyn Bell, they dubbed it LGM-1 (Little Green Men 1). It was thought nothing natural could beat out such an unwavering rhythm. Today, we know they are the most precise timekeepers Nature has to offer. So much so that astronomers have talked about using them as the basis for forms of galactic Internet and GPS. We've also used them several times to indicate our location in the galaxy to any potential intelligent civilizations.

GAMMA-RAY BURSTS

If you thought supernovas were powerful, they're nothing compared to the fury of a gamma-ray burst (GRB). They can give out more energy in a short, forty-second salvo than the sun will emit in its 10-billion-year existence and can be seen for billions of light years across the universe as bright points of light. They were first discovered in 1967 by Cold War satellites designed to pick up covert tests of nuclear weapons.

GRBs fall into two categories: short (under two seconds) and long. They're still largely a mystery, but long GRBs are believed to result from massive stars detonating as supernovas. Short GRBs – which make up about 30 per cent of all GRBs – probably come from two neutron stars colliding.

Thankfully, all GRBs spotted so far have been in the distant universe. However, a GRB flashing through our solar system could be catastrophic. If Earth was caught in the beam – which is extremely unlikely – our ozone layer would be obliterated in an extinction event (a widespread and rapid decrease in the biodiversity on Earth).

Black holes

Gravity is a very weak force. Despite 6 trillion trillion kilograms of planet beneath your feet, you can jump into the air or take off on a plane. But your freedom is only temporary – usually what goes up must come down. Unless, that is, you launch something upwards incredibly fast. If you could jump off the ground at eleven kilometres per second you would get away from Earth's gravity before it could pull you back down. This *escape velocity* is what rocket scientists must achieve if they want to get their payloads into orbit.

The more massive and compact an object, the higher its escape velocity. Jupiter, the sun, white dwarfs and neutron stars have increasingly larger escape velocities. However, the collapsing cores of the biggest stars create an object so dense that its escape velocity exceeds the speed of light. As nothing can travel through space faster than light (see page 46), nothing can escape these black holes. That's where they get their name – they are black because all light falling on them is sucked in.

If you venture too close, you're forever trapped by a black hole's gravity. No amount of rocket power will free you from its clutches. The point of no return is known as the event horizon. You wouldn't notice anything special as you pass through it, but doing so seals your fate. Let's say you go in feet first. The black hole will pull more strongly on your feet than on your head. Eventually the difference will exceed the strength of the atomic bonds holding you together. You'll be ripped apart. Physicists call this process *spaghettification*.

Where do your spaghettified bits end up? That's one of the thorniest questions in modern physics. Follow Einstein's General Theory of Relativity to the letter and it says the star's core collapses to an infinitely small, infinitely heavy point called a *singularity*. Time and space literally stop there. Material falling in is said to be added to the singularity.

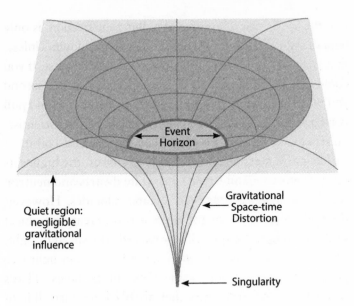

When the most massive stars die they warp space-time to such extremes that black holes form – objects from which nothing can escape.

However, this is unlikely to be the whole story as it neglects to factor in the rules of quantum physics that describe the behaviour of the universe on the smallest scales (see page 147)

Gravitational waves

The date 14 September 2015 will go down as a landmark day in the history of science. It was when we opened an unprecedented window onto the universe. But the story starts in a galaxy far, far away.

Around 1.3 billion years ago, two black holes – each thirty times the mass of the sun – smacked into one another after a dizzying inward death spiral. It was so cataclysmic that huge waves roared outwards through the very fabric of space-time itself. Hurtling away at the speed of light, these gravitational waves eventually made it to the Earth in September 2015. Fortunately, we'd just switched on a machine capable of detecting them. More gravitational wave detections from other black hole mergers followed in December 2015, January 2017 and August 2017. Waves from the merger of two neutron stars were also seen in August 2017. Soon, the trickle will become a flood.

The initial discovery was a century in the making. Albert Einstein predicted the existence of gravitational waves in 1915 as part of his General Theory of Relativity. It took us so long to find them because they are tiny by the time they reach Earth. They peter out like ripples from a stone dropped into a pond and 1.3 billion light years is a very long way to travel.

They were picked up by the Laser Interferometer Gravitational-Wave Observatory (LIGO) – two detectors in Washington state and Louisiana. Each instrument consists of two identical 4-kilometre-long empty tubes at right angles to each other. Lasers are fired down the tubes to hit mirrors at the ends. Usually the reflected laser beams get back to the start at exactly the same time.

However, if a gravitational wave passes through during the beams' passage, the space in one of the tubes is slightly stretched and contracted (gravitational waves are disturbances in the fabric of space-time itself). This means one laser beam beats the other one home.

LIGO is so sensitive that it can detect a change in the distance to the mirrors equivalent to 1/10,000th the width of a proton (the positively charged particles in the centre of atoms). That's a billionth of a billionth of a metre. Put another way, it's the same as seeing the 40-trillion-kilometre distance to Proxima Centauri (our nearest star after the sun) change by the width of a human hair.

In October 2017, three physicists behind the discovery were awarded the Nobel Prize in Physics. These detections are completely revolutionary because there are some events in the universe that only emit gravitational waves. Our eyes are now open to them for the very first time.

Time dilation

We've known since the Eddington eclipse of 1919 that massive objects warp the fabric of space around them according to Einstein's General Theory of Relativity (see page 46). Gravitational waves back up this idea even further.

But it isn't just space that's warped – time is, too. Remember that Einstein said that space and time are wrapped up together in a four-dimensional fabric called space-time. That means that time runs at different rates depending on how warped your local space-time is. Get up close to a heavy object and your time will run more slowly compared to somebody else's further away.

Even on Earth, this *time dilation* is important. Highly accurate atomic clocks stored on different shelves in a laboratory will lose synchronicity if one is closer to the ground. We also correct the clocks on board GPS satellites because time runs faster further from Earth's surface where space-time is less warped.

However, near a black hole, this effect would be far more obvious. In the blockbuster film *Interstellar*, astronauts orbiting a black hole experience one hour for every seven years that pass on Earth.

Watch someone approach a black hole and you'll see everything happen to them in ever-increasing slow motion. Eventually they would appear to freeze as they're about to cross over the event horizon. From your perspective their time has slowed to a complete stop. From their perspective, yours has.

This is gravitational time dilation, but there is another type related to your speed. You wouldn't be surprised if I said that Usain Bolt beat you in a 100-metre race. He got through space more quickly than you because he was travelling faster than you. Yet you'd probably find it odd if I said he would also get through time more quickly than you. But he would because really you're both racing through space-*time*. In this example the difference is so tiny that you're never going to notice. However, a much greater difference in speed creates a more noticeable effect.

Cosmonaut Gennady Padalka holds the record for the most number of days orbiting the Earth – 879 in total on both *Mir* and the International Space Station between 1998 and 2015. He was travelling at speeds of 28,000 kilometres per hour. Accounting for both forms of time dilation, he is 0.02 seconds younger than he would have been had he stayed on the ground. That makes him the greatest time traveller in human history, having travelled a fiftieth of a second into his own future.

White holes and wormholes

If a black hole is something you can never escape from, a white hole is a region of space you can never return to. Black holes are entrance only, white holes are exit only. At the moment they're entirely theoretical, a mathematical possibility that only exists in the equations of Einstein's General Theory of Relativity.

They appear when physicists consider what happens to material approaching the singularity inside a spinning black hole. New Zealand physicist Roy Kerr showed in the 1960s that the singularity inside a spinning black hole isn't a single point, but a ring. Normally something hitting a singularity is erased from space and time, but with a Kerr ring you might be able to pass through the middle unscathed.

Where would you end up? Kerr's solutions to Einstein's equations suggest you would pass through a tunnel known as an Einstein–Rosen bridge, before being spat out of a white hole on the other side. Some say you emerge in another part of our universe, others in another universe entirely. Either way, as white holes are exit only, you cannot retrace your steps.

An Einstein–Rosen bridge often goes by a more colloquial name: a *wormhole*. The name comes from the choices a worm has when navigating an apple. It can either crawl round the outside or chew a shorter path through the middle. Wormholes have been extensively deployed in science fiction as shortcuts in both space and time. Indeed, the physics of wormholes suggests it may be possible to use them to travel to the past. But, if they exist – and that's a big if – they're likely to be very unstable and to close very quickly.

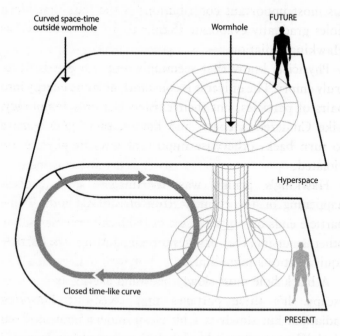

Curved space-time
outside wormhole

Wormhole

FUTURE

Hyperspace

Closed time-like curve

PRESENT

It might be possible for space-time to be warped in such a
way that shortcuts are created. If so, we may be able to use
them to travel though time, too.

So, for now, both white holes and wormholes are just
mathematical curiosities, although that may change if we
can find a Theory of Everything (see page 147).

Hawking radiation

Theoretical physicist and cosmologist Professor
Stephen Hawking has spent his entire professional
life pondering the weird nature of black holes. One of

his most important contributions is the idea that black holes gradually evaporate thanks to an effect known as Hawking radiation.

Physicists know that seemingly empty space is never truly empty. The universe is constantly turning energy into pairs of particles, but their existence is only temporary. Like Cinderella's coach and horses, they quickly have to turn back, otherwise important laws of physics are violated.

Hawking's genius was to imagine this process happening on the event horizon of a black hole. If one particle ends up on the inside of the black hole, while the other is outside, they can never turn back into the particle equivalent of a pumpkin.

A black hole must slowly be losing energy as particles escape. It's these particles that constitute Hawking radiation. But slowly is a bit of an understatement. Even glacial doesn't cut it. A black hole the mass of the sun would take around 20 million quadrillion quadrillion quadrillion quadrillion years to evaporate completely. That's 2 with 67 zeroes after it!

Nevertheless, it means black holes aren't entirely black – they glow slightly with Hawking radiation.

The Theory of Everything

Stephen Hawking's work on the evaporation of black holes via Hawking radiation uses the two biggest theories in physics: quantum mechanics – the rules of particles on the smallest scales – and Einstein's general relativity.

Black holes are a unique environment where both of these regimes are important. Normally there's no need to worry about quantum physics if you're looking at gravity and the orbits of planets. Equally, you don't bother with gravity when explaining how atoms work. A black hole is different. As a star collapses, a huge amount of material is crammed into a very small space. Suddenly gravity matters on the atomic scale.

General relativity describes how gravity is the result of warped space-time, and if you follow its rules to the letter then a black hole warps space-time into something called a singularity – an infinitely small, infinitely heavy point where the notions of space and time cease to exist. But what does it really mean for something to be infinitely small or heavy? And surely the rules of quantum physics have something to say about a region smaller than an atom?

Physicists are well aware of these issues and have been trying to unite quantum physics and general relativity together into a single theory: a one-size-fits-all framework for explaining everything in the universe from the tiniest subatomic particle to the grandest galactic supercluster. It is called a Theory of Everything (TOE).

Yet this quest has seen physicists frustrated at every turn. The two theories just won't play nicely together. They're utterly incompatible, with the application of one to the other creating irreconcilable differences. This has pushed physicists to explore extreme possibilities, including that there are more dimensions than the three of space and one of time that we're used to.

(Super)String theory and Loop Quantum Gravity

String theory has become part of popular culture in recent years thanks to Sheldon Cooper, the socially unaware genius in the hit CBS show *The Big Bang Theory*. It's one of the ways physicists are trying to unite quantum physics and gravity into a Theory of Everything.

The basic premise is that everything we see around us is ultimately made out of tiny vibrating strings. Just as you can play strings on a musical instrument in different ways to create different notes, different string vibrations create different subatomic particles. When this picture was merged with a theory called supersymmetry (see page 172), it became known as superstring theory.

Using this model, string theorists are able to combine quantum physics and general relativity mathematically, but the equations only work if there are nine dimensions of space. To explain why we only experience a three-dimensional world, physicists argue the others must be curled up so incredibly small that they remain out of sight. However, there is currently no evidence that these extra dimensions exist or that superstring theory is anything more than an elegant mathematical mirage.

In early series of *The Big Bang Theory*, Sheldon's nemesis is Leslie Winkle, a rival physicist working on Loop Quantum Gravity. She's trying to solve the problem of combining quantum physics and gravity from another angle.

Einstein said space-time is a continuous fabric warped by massive objects to create the effect of gravity. But in quantum physics nothing is continuous. Loop Quantum Gravity seeks to make space-time quantum by saying it isn't continuous either, but made of closed loops knitted together to form a fabric. It's a bit like a duvet cover. At first it looks like one continuous fabric, but under a microscope you'll see it's really made of individual stitches.

In this picture space-time isn't perfectly smooth, but grainy. That's something that is potentially testable. Astronomers are looking to see if light travelling to us from far-off galaxies has been changed along the way by this underlying structure.

Exoplanets

The habitable zone

Thanks to our fleet of orbiting satellites we have spectacular pictures of the Earth from on high. Some of the most striking images are those taken at night, when the planet's sprawling metropolises shine out as beacons of civilization. It is very clear our world is dominated by a technological species.

A closer look at the area south of the Mediterranean is particularly telling. The dry and barren lands of North Africa sit juxtaposed to the busyness of Europe, with very few electric lights on show. Yet there's one area in the north-eastern corner of the continent that's lit up like a Christmas tree: the Nile delta. In a region where water is hard to come by, people have flocked to the banks of the world's longest river.

It's a clear reminder of the importance of water to life on Earth. Life has colonized our planet in a big way, from deep underground to high up in the clouds. Yet every life form discovered so far is reliant upon liquid H_2O for survival. So, naturally, water is at the forefront of astronomers' minds when they talk about the chances of finding life elsewhere in space.

Earth sits in the *habitable zone* – the narrow region around a star where temperatures permit its existence. Too close and water boils; too far away and it freezes. This explains why the habitable zone often goes by another name: the *Goldilocks Zone*. Like the porridge in the fairy tale, it's not too hot, or too cold, but just right. Astronomers are currently searching for signs of extra-terrestrial life by looking for planets in the habitable zones of other stars.

But it's not the only place to look. In our own solar system liquid water is thought to exist in the subsurface oceans of

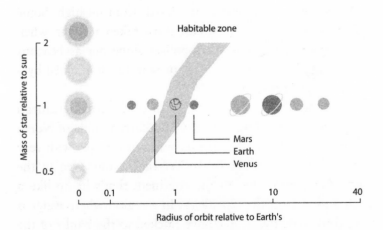

The habitable zone is a narrow region around a star where temperatures are right for liquid water. Its exact location depends on the temperature of the star.

the moons Europa and Enceladus and they are nowhere near the traditional habitable zone. Their heat comes from tidal interactions with Jupiter and Saturn instead. So it is a good place to start, but we shouldn't restrict our searches for life exclusively to the habitable zone of stars.

RED DWARF HABITABILITY

Habitable zones are moveable feasts – their location depends on the star. Around the hottest O and B stars, planets need to be considerably further away to prevent water boiling. Planets orbiting the coolest K and M stars – the red dwarfs – need to huddle closer in to stay warm.

This proximity might be a problem as the habitable zone is within the tidal lock radius – like the moon around the Earth, a planet that close will always show the same face to its star. One side is baked while the other shivers. Red dwarfs also emit powerful stellar flares and intense UV radiation – both are threats to biology.

The stakes are high because red dwarfs make up around 75 per cent of stars – a lot of potentially habitable real estate to chalk off. Recently, astronomers have used computers to model the atmospheres of these planets and that has provided some hope. They have indicated that winds could spread the star's heat more evenly around the planet, making it a less extreme environment for life.

The transit method

F inding planets around other stars – called *exoplanets* – is no easy task. Let's flip our perspective and imagine there's an alien civilization out there trying to see if the sun has any potentially habitable planets. For starters, the sun is a million times bigger than the Earth. It also shines with a fierce light, whereas the Earth creates no natural illumination of its own. The problem is compounded when you realize that the nearest they could be searching from is the closest star to us after the sun. That's Proxima Centauri some 40 trillion kilometres (4.2 light years) away. Searching for alien planets is akin to looking for a small, dark needle in a giant, blinding haystack placed so far away you can barely see the haystack let alone the needle.

These challenges have forced astronomers to invent clever ways of teasing out the presence of an exoplanet even though they cannot be observed directly. One of the leading techniques is called the *transit method*. If an exoplanet passes directly between its star and us – a transit – it will block out some light and cause the star to get temporarily dimmer.

From this simple idea – that planets sometimes block starlight – we can learn a huge amount about the exoplanet. The bigger the planet, the more light it will block and the more the star will dim. If we see multiple, evenly spaced transits, then the gap between them tells us how long it takes the planet to orbit the star. The longer it takes to go round, the further away it orbits, so we can use that to say whether or not it's in the habitable zone.

Since 2009, NASA's Kepler space telescope has been flitting back and forth between hundreds of thousands of stars looking for dips in brightness caused by transiting exoplanets. It has revolutionized our understanding of what's out there. So far it has found over 2,000 alien worlds, with some orbiting in the habitable zones of their stars (see 'What have we found so far?' on page 156).

The radial velocity method

Not all exoplanets give away their presence through a transit. If the planet doesn't cross the direct line between us and the star we won't see any changes in the star's brightness. Imagine looking down on the north pole of the sun – none of its eight confirmed planets would be seen transiting from that angle.

However, planets do have another observable effect on stars. We're used to thinking of the sun having a gravitational pull on the planets, but planets also pull on their stars. The pull of Jupiter and Saturn in particular causes the sun to wobble slightly. This wobble causes changes in the light we see from stars, known as the *Doppler effect*.

The Doppler effect with sound is something very familiar. As an ambulance speeds towards you its siren has one pitch, but as it hurtles past its sound very clearly changes. That's because the sound waves are being bunched up as the ambulance approaches, then stretched out as it moves away. Light also behaves as a wave, except it isn't the pitch that changes: it's the colour. Light sources moving away from us appear redder (*redshift*), approaching sources turn bluer (*blueshift*).

GRAVITATIONAL MICROLENSING

Massive objects bend light around them according to Einstein's Theory of General Relativity. It's what Eddington confirmed with his photographs of the 1919 eclipse (see page 49). When a massive object passes in front of a star, it magnifies the distant star's light just a like a lens. This is known as *gravitational microlensing*. The magnification is very symmetrical if the foreground object (lens) is a single object such as a star. The background object grows brighter over several weeks and fades away in the same amount of time. But if the star is accompanied by a planet then you sometimes get a spike in magnification where the planet contributes some lensing of its own. It's a bit like the main lens has an imperfection in it.

Gravitational microlensing is better for finding planets reasonably far from their stars. That complements the radial velocity and transit methods, which are biased towards nearer planets that cause more noticeable wobbles or bigger dips in their star's brightness.

Here's how it works in practice. Astronomers split starlight using a spectrometer in order to see the black, barcode-like absorption lines that help them age a star (see page 131). Imagine an exoplanet is causing its star to wobble towards us and then away from us in a repeating cycle. The star's absorption lines will constantly shift back and forth too, first towards the blue end of the spectrum, then the red.

This technique – called the *radial velocity method* – is now so sensitive that it can detect changes in a star's speed as small as one metre per second. Think about that for a moment. From a distance of hundreds of trillions of kilometres away we can spot changes in a star's speed equivalent to walking pace.

These radial velocity measurements tell us how massive a planet is. The heavier the planet, the more its star wobbles and the further the black lines swing back and forth.

What have we found so far?

You pull back the curtains just in time to watch the second sunrise of the day. As you step outside you're followed across the plains not by one shadow, but two. By day's end, one sun chases the other over the horizon. This unusual panorama is what any inhabitants of the planet Kepler-16b might experience.

Discovered in 2011, it was the first undeniable example of a circumbinary planet – one that orbits two stars. Along with the dual sunsets, sunrises and shadows, your two suns would eclipse each other every three weeks. It would be quite a show.

Kepler-16b is just one of the thousands of exoplanets we've found around stars since the first was discovered in 1995. Initially we thought we might find many copies of our own solar system, but instead we've been forced to confront the prospect that neighbourhoods like ours may be rare.

Some of the first exoplanets to be discovered were so-called 'hot Jupiters' – massive planets that orbit their stars in a matter of hours and where temperatures are high enough to melt rock. Other planets have wildly swinging temperatures due to their highly elliptical orbits. On HD 80606b, the temperature leaps from 800K to 1,500K in just six hours during its closest approach. The Kepler-11 system has six planets, with five orbiting closer to their star than Mercury does to the sun. The planet 55 Cancri e might even have a surface covered in diamonds, formed in its hot, pressurized interior.

Naturally, however, the exoplanets garnering the most attention are those with the potential to be most like our own. In 2014, astronomers found Kepler-186f – the first Earth-sized planet discovered in the habitable zone of its star. That was followed a year later by Kepler-452b. By 2017 astronomers were announcing seven Earth-sized exoplanets around the star TRAPPIST-1, three of them in the habitable zone. There's even a potentially habitable planet around Proxima Centauri, the nearest star to us after the sun.

But 'potentially habitable' comes with a huge pinch of salt. All astronomers are really saying is that *if* the planet has the same atmospheric composition as the Earth then the temperature is probably suitable for liquid water. They're not saying it is guaranteed to be inhabited, or even that it has water. So their next job is to measure the atmospheric compositions of exoplanets and prove they have water.

SUPER-EARTHS

Our solar system has small rocky planets and giant gas planets. There's no intermediate planet (save for maybe Planet Nine). One of the biggest surprises to emerge from exoplanet research is the discovery of a new class of planet: the super-Earth.

Rocky, but with masses several times greater than ours, these super-Earths have much stronger gravity. There's much debate as to whether this is a help or hindrance for life.

Any land would be much flatter, with mountains unable to form as high as they do on Earth. The Earth's surface is roughly 70 per cent water and 30 per cent land, but super-Earths might be true water worlds with only a tiny fraction of land above sea level. The biggest might be completely submerged. More massive planets would have hotter, larger cores leading to stronger magnetic fields offering greater protection from dangerous solar activity and cosmic rays.

Stronger gravity also means the ability to cling on to more gas and hence a thicker atmosphere. That's a boon for astronomers, because more substantial atmospheres are easier to characterize.

Characterizing atmospheres

Right now you're breathing in an atmosphere that's 21 per cent oxygen. Even sitting around doing nothing you consume 550 litres of the stuff every day. Over your lifetime that's more than 16 million litres or 22 tonnes of oxygen.

The trouble is oxygen shouldn't be here. It is a very reactive gas and it quickly combines with other elements in the atmosphere to create new chemical compounds. And yet there's plenty for you – and everyone else – to breathe in. We have other life forms to thank for that. Plants, trees and microbes in the ocean produce oxygen via photosynthesis to replace what's being lost.

That makes oxygen a biosignature gas – one that if seen in abundance may indicate the presence of life on a planet. Astronomers would dearly love to look for biosignature gases in the atmospheres of some of the Earth-sized exoplanets they've been finding in the habitable zones of stars. But that's no mean feat.

The good news is that the process has already been road-tested on much larger exoplanets, particularly the hot Jupiters, whose atmospheres are puffed up by the extreme heat they experience. In 2017, astronomers even measured the atmosphere of the super-Earth GJ 1132b – a planet only 40 per cent bigger than ours. Telescopes capable of exploring the atmospheres of Earth-sized planets are currently under construction and will soon see active service.

They'll use the same technique astronomers deploy to discover what stars are made of: spectroscopy (see page 50).

As an exoplanet transits in front of its host star, some starlight will pass through its atmosphere and continue its journey to our telescopes. But some colours won't make it out of the atmosphere because chemical compounds there will swallow that wavelength of light. The resulting spectrum will contain black absorption lines telling us what the atmosphere is made of. Along with oxygen, we're looking for signs of water and other potential biosignature gases such as methane.

Exomoons

Most of our attention to date has been on exoplanets. Quite rightly, too – it is the logical first step, as the only life we know about started on a planet. Yet science-fiction writers have long considered the possibility that life can exist on a moon orbiting around a planet. In *Avatar*, the action is set on Pandora – a lush, rocky moon orbiting the gas planet Polyphemus. In *Star Wars*, Endor is a forest moon home to the Ewoks. In *Doctor Who*, the Doctor considers retiring to the Lost Moon of Poosh, famous for its swimming pools.

A star without rocky planets in the habitable zone might still have worlds capable of sustaining life. Drag Jupiter into the sun's habitable zone and conditions on some of its planet-sized moons might well be plush. But if finding exoplanets is a tall order, spotting exomoons is really pushing the boundaries of what we're currently capable of.

That hasn't stopped a team led by David Kipping at Columbia University in New York from looking. The gravitational pull of a moon would periodically speed up and slow down a planet as it orbits its star. This would lead to transits happening up to five minutes earlier or later than expected if the planet orbited alone. Finding these clues is incredibly precise work and right at the limit of what the Kepler space telescope can achieve. It would take your desktop computer fifty years to churn through the calculations required to check just one planet.

Nevertheless, the astronomical community was buzzing in the summer of 2017 with rumours of a potential exomoon in the Kepler-1625b system. It looks like there could be a Neptune-sized moon in lockstep with a Jupiter-sized exoplanet. At the time of writing, Kipping and his team have applied for time on the Hubble Space Telescope to take a closer look in the hope of confirming what would be a historic discovery.

CHAPTER 5

Galaxies

The Milky Way

Name and appearance

The stories of the Native American Cherokees call it 'The Way the Dog Ran Away', a trail of stolen cornmeal dropped by a thieving canine. In East Asia it is the silvery river of heaven. The Māori of New Zealand see a giant canoe. In Greco-Roman mythology it is the breast milk of Hera sprayed across the sky by a young suckling Heracles (Hercules).

From this last story we get our modern scientific name for the bright, dusty arch that stretches from one side of the night sky to the other: the Milky Way. A band of light approximately 30 degrees across, it is dominated by star clusters and dark dust lanes. Yet many of us have never seen it. Eighty per cent of people in North America live in areas where it is obscured by light pollution. About a third of the world's population are in the same boat.

It's well worth venturing out into a rural, dark spot in order to see it for yourself. It is arguably the greatest spectacle the night sky has to offer. Galileo was the first to run a telescope along it, seeing innumerable stars. Even

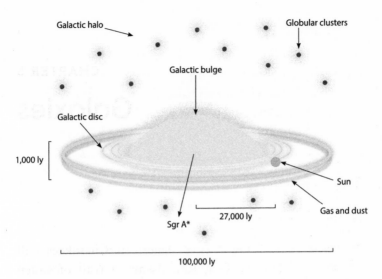

Our galaxy – the Milky Way – is flat and disc-shaped with a central
bulge and spiral arms, all embedded in a halo of dark matter.

a pair of binoculars will reveal a sky teeming with stars
and dust. The Great Rift and Coalsack are conspicuous
dark regions devoid of stars. Giant molecular clouds are
blocking our view of the stars that lie beyond.

Although the Milky Way is visible the world over,
the real action is centred on the zodiac constellations of
Sagittarius and Scorpius. The best views of this area are
from latitudes of –30 degrees because it is directly overhead.
That line runs from Chile and Argentina, across to South
Africa, before continuing eastwards passing close to the
Australian cities of Perth and Brisbane. It is no surprise
that some of the best telescopes in the world are built near
to this line. Astronomers want a front-row seat to study
the Milky Way and its mysteries.

Shape, size and contents

We see the Milky Way as we do because we live inside it. A spiral galaxy, from the outside its shape would resemble two fried eggs stuck back to back. There's a yolk-like bulge in the centre surrounded by a much flatter disc. We live about halfway out in that disc on one of the galaxy's minor spiral arms.

When we look towards Sagittarius we are peering directly through the disc to the crowded central region. Journey towards the constellations Orion and Auriga and you're heading in the opposite direction to the galaxy's edge.

Estimates of the size and contents of our Milky Way vary considerably. However, astronomers agree that the galaxy is at least 100,000 light years wide. That's a whopping 1 million trillion kilometres. A beam of light that set off from one side of the galaxy 100,000 years ago, back when *Homo sapiens* still shared the planet with Neanderthals, would only now be making it to the other side.

Here's another way to picture it. Imagine that you could shrink the distance from the sun to the edge of the Kuiper Belt down to the size of your little finger. On this scale the Milky Way would stretch across the Atlantic Ocean, with one edge in London and the other in Kingston, Jamaica. The sun is tiny compared to the galaxy at large.

It may be very wide, but on average the disc of the Milky Way is just 1,000 light years thick. That disc is home to the sun and at least 100 billion other stars, perhaps as many as 400 billion. Estimates based on Kepler space telescope data suggest there could be 60 billion planets in the habitable zones of these stars.

The stars in the disc are rotating around the centre anti-clockwise – that's in the same direction the planets orbit the sun. It takes the sun approximately 220 million years to complete one lap of the Milky Way, a period astronomers call a *cosmic year*.

Spiral arms

We cannot see the Milky Way from the outside – it is just too big to leave. At the speed of the *Voyager* probes it would take 5 million years to get out via the shortest route. However, if we could, our galaxy's most striking feature would be its spiral arms.

Four vast chains of stars and gas appear to curl outwards from a central bar located in the galactic bulge. They're joined by at least two smaller arms, one of which is home to the sun. We're able to build up this picture by looking at how the stars in our galaxy move, as well as looking at other spiral galaxies in the wider universe.

For many years spiral arms were a conundrum. At first glance it looks like each arm is a single group of stars moving together around the centre. But that can't be right. Spiral galaxies rotate reasonably quickly and so the arms would wind up over time. Picture the galaxy as a running track with several lanes. Just like runners on the inside track, stars closer to the middle would get ahead of those further out. It would only take a few orbits for the arms to disappear.

In the 1960s, Chinese astronomers C. C. Lin and Frank Shu realized spiral arms are more like traffic jams. When someone brakes, everyone behind them brakes too. As the offending car accelerates away, the jam moves backwards through the traffic like a wave. When you encounter this overly dense region of cars, you're going to slow down. It's the same with stars. If molecular clouds also get compressed it will trigger their collapse into new stars (see page 132). That explains why we see so much star formation in spiral arms.

These traffic jams – which astronomers call *density waves* – have one aspect you won't encounter on any motorway. As a star approaches a dense region, it gets pulled into it quicker by the collective gravity of the stars already in the jam. When the star eventually makes it out, it does so very slowly because the gravity of the stars behind holds it back. So stars spend a long time as part of a density wave and spiral arms persist.

The galactic centre

The sun sets over a dormant Hawaiian volcano as the domes of the giant Keck Observatory are thrown into silhouette. They slowly open as the curtain of night falls, revealing the ten-metre mirrors inside. Since the mid-1990s, astronomers have been using these telescopes atop Mauna Kea to collect ancient light falling to Earth from the centre of the Milky Way.

Their quest has been to understand exactly what everything in the galaxy is orbiting. Peering through 27,000 light years of gas and dust, they've spotted stars whizzing around a bright source of radio waves known as Sgr A* (said 'Sagittarius A Star'). We can use the stars' speed and distance from Sgr A* to calculate the mass of the object they're orbiting. It tips the scales at a colossal 4 million suns. For the stars to be in stable orbits it also has to be smaller than 12 million kilometres across (about a fifth of the Mercury–sun distance or 8.5 times the diameter of the sun). The only thing capable of cramming so much mass into a relatively small space is a supermassive black hole.

So, right now, the sun is dragging us around a black hole at speeds of nearly a million kilometres per hour. Fortunately, we are far enough away not to be sucked in, but astronomers have seen material get perilously close. Between 2011 and 2014 they watched a gas cloud called Sagittarius G2 skirt around the black hole. At first they thought it was going to disappear into the void, but it seems there was a star inside the cloud that helped hold it together.

Another cloud – Sagittarius B2 – has been hit by a flood of radiation produced by the black hole around 400 years ago. It suggests that, relatively recently in cosmic terms, Sgr A* has been a million times more active than it is now.

The Event Horizon Telescope

The galactic centre is the perfect laboratory to test Einstein's General Theory of Relativity. The stars orbiting around Sgr A* experience a gravitational pull one hundred times greater than anywhere the theory has been tested so far. Just as Mercury's proximity to the sun showed us the flaws in the Newtonian picture of gravity (see page 48), so the stars around the Milky Way's central black hole could poke holes in Einsteinian ideas. Any deviations might point the way to a successful Theory of Everything (see page 147).

Continuing observations of the central stars with the Keck telescopes will go a long way to achieving this. However, we need to look even closer to the black hole if we really want to put Einstein in the spotlight. Ideally, we want to see how space-time is warped just outside the event horizon (see page 143).

General relativity predicts that a black hole should have a circular shadow: a dark patch inside a ring formed by light that is initially moving away from you but is bent back around towards you by the black hole's extreme gravity. If it turns out not to be circular, or a different size to what we expect, then we might have a revolution on our hands.

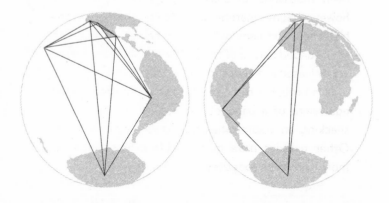

You need a telescope the size of Earth to see the area around a black hole. So astronomers made the Event Horizon Telescope by linking existing telescopes across the world.

THE FERMI BUBBLES

Our galaxy still throws up surprises. In 2010, astronomers using the Fermi Space Telescope (named in honour of pioneer Enrico Fermi – see also page 179) found two vast gamma-ray bubbles blowing away from the galactic centre. Inflating above and below the disc, these Fermi Bubbles extend for 25,000 light years in each direction. Each contains enough cool gas to create 2 million suns. Astronomers believe they formed between 6 and 9 million years ago, a heartbeat in the history of a galaxy as old as the Milky Way. Their formation was triggered by Sgr A* consuming a gas cloud weighing the equivalent of hundreds, maybe even thousands, of suns. Not everything went in, however. Some material was accelerated around the black hole and back out into the galaxy. The smooth and rounded nature of the bubbles implies this energy was released in a very short amount of time.

It is probably the last time Sgr A* gorged on the equivalent of a banquet – it appears to have been snacking on a somewhat restricted diet ever since. Other galaxies have monster black holes with far more voracious appetites.

Yet focusing on such a compact object 27,000 light years away is no mean feat. It requires a telescope with a resolution two thousand times greater than the Hubble Space Telescope. A single telescope of that power is out of the question – it would need to be the size of the Earth.

So astronomers have devised a clever alternative. They've linked up existing telescopes in the US, Mexico, Chile, Antarctica and Spain, and together they mimic a telescope almost as wide as the planet. In 2017, this Event Horizon Telescope took its first look at Sgr A* and it will soon be putting Einstein's theories to the ultimate test.

The rotation problem

On the face of it, the Milky Way looks like a giant solar system. There's a big mass in the middle and lots of smaller objects orbiting around it. But a closer inspection reveals that spiral galaxies are fundamentally different to planetary systems.

Work your way out from the sun and each planet moves more slowly than the last, taking longer to go round. Mercury takes just eighty-eight days, Neptune takes 165 years. So you might expect the orbital speeds of stars to drop off also as you head away from the galactic centre. They don't.

The first clues that the Milky Way had this rotation problem came as early as the 1930s. Jan Oort – the Dutch astronomer who gives his name to the Oort Cloud – was looking at stars near the edge of the galaxy and measuring their speed. He found that they were moving too fast. At the speed they were going they should have escaped from the gravity of the Milky Way and been lost to intergalactic space. The fact that they weren't led him to suggest the gravity of the galaxy must be stronger than we thought.

Oort's work was largely forgotten until the late 1960s when the baton was picked up by American astronomer

Vera Rubin. Over the next decade she looked at the rotation of a hundred other spiral galaxies and found the same phenomenon. Stars on the edges of spiral galaxies orbit just as fast as those closer to the bulge. Rubin died on Christmas Day 2016 amid suggestions she would have received the Nobel Prize for her work (prizes aren't awarded posthumously).

Dark matter

The most popular explanation for the rotation problem is that there's extra, invisible mass hidden throughout the galaxy that we cannot see. This *dark matter* would provide the additional gravity needed to keep hold of the fast-moving stars. Indeed, Oort suggested in the 1930s that the hidden stuff might outnumber the visible material three to one.

An early frontrunner was the idea that the galaxy contains lots of Massive Astrophysical Compact Halo Objects (MACHOs). Astronomers love their acronyms. Essentially, MACHOs are normal objects such as black holes and neutrons stars that are so physically small that they're hard to see, but so heavy that together they could make up the shortfall in gravity.

Except that we can measure the mass deficit a lot more accurately today than Oort was able to. The stuff we can see only makes up 10–12 per cent of the Milky Way's mass. That's too big a gap for MACHOs to fill on their own. We can occasionally spot a MACHO if it passes in front of a distant star and magnifies its light through gravitational microlensing (see page 155). We haven't seen enough of these events to suggest there's a population of MACHOs remotely big enough to solve the rotation problem.

So astronomers now think dark matter may instead come in the form of WIMPs (Weakly Interacting Massive Particles). Weakly interacting because they don't interact with light (hence why we don't see them). Massive because they have to make up for the significant shortfall in gravity. Unlike MACHOs, WIMPs are something we've never encountered before. They're an entirely new type of matter dreamt up by particle physicists to explain the rotation of galaxies.

Everything we see around us is made of particles from the Standard Model – it's like a cookbook for the cosmos developed incrementally by particle physicists over many decades. Except that no ingredient in the Standard Model behaves like dark matter. However, physicists have been working on an extension to the Standard Model called supersymmetry (we encountered it when we looked at superstring theory on page 149). It says that every particle in the Standard Model has a mirror particle. WIMPs could be the lightest of these supersymmetric particles – the *neutralino*.

The search for WIMPs

In an abandoned goldmine 1.5 kilometres under South Dakota sits a tank of liquid xenon shielded by 70,000 gallons of water. Meanwhile, in Antarctica, detectors deep under the ice are ready for action. At the Large Hadron Collider in Switzerland, particles smash together at close to light speed. Above the Earth, the AMS-02 experiment orbits every ninety-two minutes strapped to the International

Space Station (ISS). Physicists are using all these instruments to search for the universe's most wanted: WIMPs. If dark matter really is an artefact of supersymmetry, then particle physicists at CERN need to find evidence that supersymmetry is more than just a nice theory on paper.

If WIMPs really exist, one hits your body every single minute. However, detecting them when there's so much else going on around you is as good as impossible. So in the South Dakotan goldmine, the Large Underground Xenon (LUX) experiment is shielded by rock and water. It is designed to pick up flashes of light caused by a stray WIMP striking the xenon.

The detectors of the IceCube experiment near the South Pole are similarly protected by the frozen tundra. They are on the hunt for indirect evidence of WIMPs. If the galaxy contains dark matter then the sun should be sweeping some up with its gravity as it orbits the Milky Way. This would mean WIMPs end up colliding with each other deep inside our star. Calculations suggest this would produce high-energy neutrinos which can make it out of the sun – that's what IceCube is looking for.

Finally, the Alpha Magnetic Spectrometer (AMS-02) currently hitching a ride on the ISS is looking towards the busy central bulge of the Milky Way. With matter more tightly packed there, WIMP collisions should be more common. These events are thought to create a cascade of particles called positrons (the antimatter equivalent of electrons). Find an excess of positrons near the galactic centre and you might have a smoking gun. Tantalizingly, a burst of positrons has been found. However, astronomers can't yet rule out less exotic explanations.

The AMS-02 experiment orbiting on the ISS is looking for a spike in positrons created by dark matter collisions in the heart of the Milky Way.

As you can see, physicists have been to extreme lengths to snare a WIMP. But so far all searches have come up shy. It's still the best idea we have, but if we don't find one soon we may have to return to the drawing board. Advocates of a completely different idea – MOND – are already sensing blood and are poised to step in.

Modified Newtonian Dynamics (MOND)

We need dark matter to explain why we don't see enough gravity in galaxies to account for the speed of their stars. So we've invented invisible stuff to make up for the shortfall.

But what if we don't really understand gravity properly? What if we don't see enough gravity because we don't really

THE GALACTIC HALO

A spiral galaxy may look flat, but that's only the visible part of it. The Milky Way appears to be embedded in a vast halo of dark matter. This halo is shaped like a squashed beach ball, with more dark matter above and below the disc than at the sides.

Astronomers have mapped this out by tracking dwarf galaxies orbiting the Milky Way. Our galaxy has around fifty of these small satellites, each with far fewer stars than a galaxy like the Milky Way (see page 180). Just as we use orbiting stars to weigh the supermassive black hole Sgr A*, we use the satellite dwarf galaxies to weigh the Milky Way.

The galactic halo is also home to many globular clusters (see page 130). These dense groups of ancient stars are spectacular through binoculars or a telescope. Up to 40 per cent of the Milky Way's globular clusters orbit retrograde – in the opposite direction to the stars in the disc. Like retrograde moons in our solar system, that probably means they're captured objects.

know how the force works on scales as big as galaxies? That's exactly what advocates of Modified Newtonian Dynamics (MOND) argue. MOND is the idea that gravity isn't the universal law envisaged by Newton – it requires modification on the biggest scales. It was first proposed in 1983 by Israeli physicist Mordehai Milgrom.

The acceleration of a typical star around a spiral galaxy is 10 billion times less than Newton's apocryphal apple experienced as it fell to Earth. Milgrom argued that we need to change Newton's equations under such tiny accelerations. MOND theorists say that objects in weak gravitational environments experience a slightly stronger pull than we'd normally expect.

For any scientific theory to be taken seriously it needs to make testable predictions. Proponents of MOND used its modified equations to predict the orbits of seventeen dwarf

THE BULLET CLUSTER

Located nearly 4 billion light years from Earth, the Bullet Cluster is actually two clusters of galaxies in the process of colliding. Astronomers have mapped out the way hot gas is spread throughout the merger. They've also used the fact the cluster bends light from distant objects – gravitational microlensing – to work out how the mass is distributed inside (see page 155).

There's a clear separation between the hot gas and the majority of the mass. Most of the mass must therefore be invisible. Many hold it up as a smoking gun for dark matter and incontrovertible proof against MOND. But in recent years MOND backers have come up with ways to explain this discrepancy, too.

Critics of dark matter also point to the speed at which the clusters collided – 3,000 kilometres per second. Such high speeds didn't crop up in early computer models of dark matter. But now the models have been tweaked to fit. So the Bullet Cluster remains a huge bone of contention.

galaxies orbiting around Andromeda (the nearest major galaxy to the Milky Way). They got them spot on.

However, MOND remains a fringe theory. Most astronomers and cosmologists are in favour of the dark matter approach. That's largely because dark matter being a physical entity helps explain how structure formed in the early universe. The gravitational attraction of dark matter helped to clump ordinary matter together into stars and galaxies in a universe expanding after the Big Bang.

It also explains why Andromeda and the Milky Way are currently on a collision course (see page 185). To have overturned the expansion of the universe, and now be heading towards each other, there needs to be a gravitational attraction between them equivalent to eighty times more matter in the galaxies than the stars we can see.

The Drake Equation

Long before they found the first exoplanet, astronomers wondered about the possibility of life elsewhere in the universe. Back in 1600, Giordano Bruno was arguing that the stars are just distant versions of the sun, with planets and perhaps living things (see page 23).

In the early 1960s, US radio astronomer Frank Drake came up with a way of estimating how many intelligent civilizations there might be in the Milky Way. He presented his work to the first meeting dedicated to the Search for Extra-Terrestrial Intelligence (SETI). As a radio astronomer, he was specifically interested in the number of civilizations it might be possible to communicate with.

The Drake Equation is an exercise in probability. To find the overall probability of two events happening you multiply their individual probabilities together. So the chances of a coin coming down tails twice in a row is ¼ (½ × ½). Drake noted seven key factors involved in whether a planet has an intelligent civilization capable of communicating via radio signals. A star has to have a planet *and* that planet has to be suitable for life *and* life has to get started there *and* it has to evolve intelligence, etc.

Drake multiplied these probabilities together to estimate the number of contactable civilizations in the Milky Way. His original answer was at least a thousand. Use more modern values and this number is significantly lower, sometimes just a handful. That might explain why we haven't found any evidence of other advanced civilizations so far, but astronomers are looking.

The Search for Extra-Terrestrial Intelligence (SETI)

Our search for alien signals using radio telescopes began in earnest in the early 1960s. In 1960, Frank Drake turned the 26-metre dish at Green Bank, West Virginia, towards the stars Tau Ceti and Epilison Eridani. He heard nothing of note.

But just as you have a range of radio stations to choose from, what frequency should astronomers be tuning into? Drake chose a frequency close to 1420 megahertz. Not only is it a quiet part of the radio spectrum, it is also between the natural frequencies of hydrogen (H) and hydroxyl (OH).

THE FERMI PARADOX

Where is everybody? This simple question is known as the Fermi Paradox. It's named after Italian-American physicist Enrico Fermi.

On the face of it, life in the universe should be quite common. There are loads of stars out there and many planets, so lots of chances for living things to emerge elsewhere in space. Given that there are also stars much older than the sun, there should be habitable planets much older than the Earth with civilizations far in advance of our own.

But if all this life is indeed out there, why haven't we seen or heard a single shred of evidence for its existence? On Earth, we've uncovered artefacts of dinosaurs and early hominid species that lived here before us. But we've never seen the equivalent archaeology in space to suggest anyone else is in the Milky Way now or has been in the past.

Some astronomers argue that's because we're the only ones in the Milky Way. Others that intelligent civilizations wipe themselves out before they have a chance to make themselves known to anyone else. Yet, despite the odds, we continue to listen patiently to the sky in search of signals from any potential neighbours, past or present.

Radio astronomers have noted that together these form water (H_2O). So this gap has been dubbed the 'water hole' – a quiet place in the radio spectrum where aliens might choose to meet and talk in the same way that animals meet at watering holes in the savannah.

Since Drake's initial work, astronomers have made a concerted effort to scan the skies at these frequencies. But even checking the nearest thousand stars in this narrow band means searching through over 242 billion possible radio channels. SETI received a significant boost in 2015 when Russian billionaire Yuri Milner threw $100 million at the problem. The resulting ten-year *Breakthrough Listen* project will be the most extensive search for alien communication to date.

Yet in six decades of listening we've never heard anything conclusively alien. There is, however, one signal that remains tantalizingly unexplained: the Wow! signal of 1977. A read-out from the Big Ear radio telescope in Ohio, it bears all the hallmarks of being alien in origin. A strong, seventy-two-second-long burst of radio transmission, it got astronomer Jerry R. Ehman so excited that he circled the relevant digits and wrote 'Wow!' in red pen alongside them. But we've never heard it again and have no way to prove it came from E.T. It could be the most historic document in human history, or it could be nothing. That is the frustration that's part and parcel of SETI.

The Local Group

The Magellanic Clouds

Portuguese explorer Ferdinand Magellan passed below the equator during his attempts to circumnavigate the Earth in the sixteenth century. There he noticed two giant clouds tattooed on the sky, wheeling through the night as the Earth rotated. Although he didn't know it at the time, he was looking at material beyond our Milky Way galaxy. We still know this duo as the Magellanic Clouds.

They are part of the Local Group – an aptly named collection of our nearest galactic neighbours – including the other dwarf galaxies orbiting the Milky Way and the Andromeda and Triangulum galaxies (see page 184). Almost exclusively visible from southern latitudes, the Magellanic Clouds are easily spotted with the unaided eye spanning the constellations of Dorado, Mensa, Tucana and Hydrus.

The Large Magellanic Cloud (LMC) is 14,000 light years in diameter and 160,000 light years away. In the night sky it appears as wide as twenty full moons. It's home to the Tarantula Nebula – the most active region of star formation anywhere in the Local Group. In 1987, a supernova exploded close to the edge of this nebula. Known as SN 1987a, it was the nearest supernova to detonate since the so-called Kepler Supernova of 1604. It was so bright it could be seen with the naked eye.

The Small Magellanic Cloud (SMC) is about half the size and 40,000 light years further away. Gravitational interactions with the LMC create the Magellanic Bridge – a trail of hydrogen gas spanning the otherwise empty space between them. A similar effect creates the Magellanic Stream between the Magellanic Clouds and the Milky Way itself. The presence of a distinctive bar at the centre of the LMC suggests it may have been a dwarf spiral galaxy since stripped of its arms by the gravitational pull of its neighbours.

Cepheid variables

In 1908, American astronomer Henrietta Swan Leavitt published one of the most important papers in the history of astronomy. It was entitled '1777 variables in the Magellanic Clouds'.

The variables in question were *Cepheid variables*. These stars expand and contract, leading to their brightness changing in a regular way. They also provide an invaluable way to measure distances in space, part of a group of astronomical tools known as *standard candles*. Astronomers use parallax to discover the distances to the nearest stars (see page 124). However, eventually stars get so far away that parallax no longer works. That's where standard candles come in.

Imagine looking at a light bulb through the window of a distant building. The further you are away, the dimmer the bulb will appear because light fades over distance. If you know the real brightness of the bulb (say 40 watts or 60 watts), you could work out how much it has faded and therefore how far away you are from the building.

We can do exactly the same in space, except that stars don't come with their real luminosity neatly written on the side. That's why Henrietta Swan Leavitt's work on Cepheid variables was so valuable. She discovered that more luminous Cepheids take longer to vary their brightness. Find a Cepheid, wait and see how long it takes to vary its brightness, and you can work out its true luminosity. Just as with the light-bulb example, it's then easy to work out how far away the star is.

As we'll see, huge strides in our understanding of the universe and its origins followed in the early decades of the twentieth century. But it's hard to see how any of them would have been possible without Henrietta Swan Leavitt and her accurate way to measure distances once parallax stops working.

The Andromeda and Triangulum galaxies

The night sky is a feast for the eyes. Meteors, comets, planets, stars, globular clusters, nebulae, double stars, there's a lot on offer. But what are the furthest objects you can see without binoculars or a telescope? The answer is the nearest galaxies to us.

Nestled among the stars in the constellation Andromeda, barely discernible to the human eye, sits a fuzzy patch of light. It's as if someone has licked their thumb, reached up and smudged the inky blackness. This is the Andromeda galaxy – the closest major galaxy to our own Milky Way. It contains a trillion stars, yet still looks like nothing more than a passing wisp of cloud. This is due to its incredible distance from us – a staggering 2.5 million light years. Even though light travels at 300,000 kilometres a *second*, it still takes light 2.5 million years to trek all the way here from Andromeda. It's no wonder we can barely see it.

When you look at Andromeda, you are seeing light that is 2.5 million years old. Modern humans weren't even on this planet when the light arriving here today originally set off. Instead, our ancestors, primates known as *Australopithecus*, were just starting to fashion stones into the first primitive tools at the dawn of the Stone Age.

For any alien life forms living in Andromeda, with powerful enough telescopes to peer at Earth from so far away, *Australopithecus* is all they would see. They wouldn't know that their descendants fashioned floating vessels out of dead trees to roam the oceans. Or that, in turn, their descendants conquered a whole new ocean – this time black, not blue – by launching giant metallic ships into the sky.

You'll often hear Andromeda referred to as the furthest thing you can see without binoculars or a telescope. Under most circumstances this is true. However, those with perfect eyesight should also be able to spot the Triangulum galaxy from a very dark site. The third largest galaxy in the Local Group, it sits 3 million light years away. It appears Triangulum is being gravitationally disrupted by Andromeda, with a stream of hydrogen between them stretching a whopping 782,000 light years.

Milkomeda

Andromeda and the Milky Way are on the move. The gap between the galaxies is closing at the rate of 100 kilometres per second and the pace is quickening. In around 4 billion years, the two giant star cities will likely collide with one another.

That may sound catastrophic, but spiral galaxies are not solid objects and so it won't be like a head-on car crash. The galactic discs will thread through each other instead, with gravity causing vast strands of stars and dust to flick outwards. Eventually they'll combine into one super-galaxy astronomers have dubbed Milkomeda. The new galaxy will be orbited by Triangulum, which Andromeda will drag in during its approach.

Computer models of this event suggest there is a 12 per cent chance the sun will be thrown out during the merger, orphaned to intergalactic space. Not that life on Earth needs to worry – the sun will have already baked this planet into a lifeless hellhole by then (see page 133). The consolation prize is that Andromeda will become an increasingly spectacular sight as it approaches. It already appears six times wider than the full moon.

Galactic mergers are pretty common throughout the universe and have been widely studied by astronomers. One of the most famous examples is the Antennae Galaxies in the constellation Corvus. The name comes from the streams of gas being ejected outwards from the central region – they look like an insect's antennae. The two galaxies collided just over a billion years ago, causing clouds of gas and dust to merge and triggering an intense period of star formation.

The other well-known galactic merger is in the Whirlpool Galaxy. The main galaxy has a companion called NGC 5195 – a dwarf galaxy that appears to have passed through the main disc 500–600 million years ago.

Distant galaxies

Clusters and superclusters

Journey away from the Local Group and you eventually come across other clusters of galaxies. Some of the nearest are the M81, M51 and M101 groups named after their biggest galaxies. They form part of the Virgo Supercluster, a colossal structure containing over one hundred galaxy groups including our own Local Group. It stretches for more than 100 million light years. The observable universe comprises about 10 million of these superclusters.

Getting a clear picture of this structure in your head can be tricky. Comparing it to the more familiar geography of Earth can help. Imagine that our solar system is your house,

THE VIRGO CLUSTER

The enormous Virgo Supercluster is named after its biggest, most central member: the Virgo Cluster. Our Local Group contains fifty to sixty galaxies, but the Virgo Cluster boasts around 2,000. Its total mass is over a million billion suns.

One of its most studied galaxies – M87 – is orbited by 12,000 globular clusters (compared to the Milky Way's 150). It has a central supermassive black hole weighing 7 billion suns. That compares to just 4 *million* suns for Sgr A* in the Milky Way.

A distinctive hot jet surges almost five thousand light years outwards from the centre of M87. Material is being accelerated to close to light speed by the gravity of the central black hole and ejected out of the galaxy. Astronomers hope to learn more about it using the Event Horizon Telescope (see page 167).

You can see M87 and many of the members of the Virgo Cluster for yourself using a small telescope. The cluster is visible in a ten-degree-wide patch of sky between the stars Denebola in Leo and Vindemiatrix in Virgo.

the sun and the planet are the rooms. The exoplanetary systems found so far by the Kepler space telescope and others would be the other houses on your street – separate, but very close by.

Zooming out from your street, next we'd see your town or city. A galaxy is just a city of stars and so the Milky Way is the equivalent of our home town in space. It even has busier downtown regions in the bulge and quieter suburbs further out where we live in the disc.

On Earth, cities are collected together to form countries. In space, galaxies form groups called clusters. Countries form huge, sprawling land masses called continents. Galactic clusters form massive groups called superclusters. Just as the world is made from continents, the observable universe is made of superclusters.

On Earth	In Space
Your house	Solar system
Your street	Exoplanets
Your town/city	Milky Way
Your country	Local Group
Your continent	Virgo Supercluster
Earth	Observable universe

Galaxy classification

Not all galaxies are spirals. M87, for example, is an elliptical galaxy. More like a rugby-ball-shaped blob of stars, it has no distinctive dust lanes or spiral arms. In contrast to flatter spirals, ellipticals also rotate very slowly.

Galaxies were originally classified using the Hubble sequence, named after the American astronomer Edwin Hubble. His classification system contains three categories of galaxy: elliptical, spiral and lenticular (flat discs with poorly defined spiral arms).

Hubble originally arranged these galaxies in a diagram shaped like a tuning fork. Many people wrongly believe that he was showing the way galaxies evolve, starting with round ellipticals that gradually spin faster and faster until they flatten out into lenticular and then later develop spiral arms. But that was never Hubble's plan and today we've confirmed that galaxies don't evolve in that way. Still, the tuning fork remains a useful way to classify galaxies.

Ellipticals are denoted by the letter E, followed by a number between 0 and 7. The higher the number, the more elliptical the galaxy. The symbol for lenticulars is S0. Spiral galaxies without a central bar are either Sa, Sb or Sc with

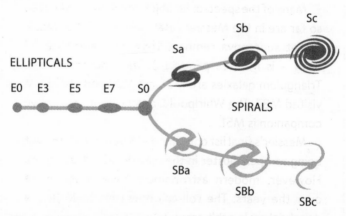

Edwin Hubble's tuning-fork diagram shows the different types of galaxies: elliptical (E), lenticular (S0) and spiral (S).

the spiral arms getting less tightly wound as you move through the alphabet. Sometimes a galaxy falls between two groups – for example Sbc. Barred spirals are SBa, SBb or SBc.

THE MESSIER CATALOGUE

Casting a telescope or binoculars around the night sky will reveal many fuzzy, cloudy patches. Some are star clusters or nebulae in our own galaxy. Others are distant galaxies like Andromeda.

The French astronomer Charles Messier compiled a catalogue of these objects in the eighteenth century. A comet hunter, his intention was to document anything that might be mistaken for a comet. Objects in the list are designated M1, M2, M3, etc.

Many of the spectacular objects we've encountered so far are in the Messier catalogue. The Crab Nebula – that supernova remnant from the guest star of 1054 – is M1 (see page 136). The Andromeda and Triangulum galaxies are M31 and M33 and we've just visited M87. The Whirlpool Galaxy with its disruptive companion is M51.

Messier's final list contained 103 items, with the last being an open cluster in the constellation Cassiopeia. However, modern astronomers have added more over the years. The roll-call now runs to M110 – a dwarf galaxy in orbit around Andromeda.

This system does have its drawbacks. How you classify a galaxy depends on the angle we're viewing it at. An old spiral galaxy that has lost the majority of its arms can easily be mistaken for an elliptical galaxy if we're seeing it face on. In 2011, the team of astronomers behind the ATLAS^{3D} galaxy survey found that two-thirds of local galaxies that had previously been classified as ellipticals were in fact fast-rotating discs.

Active Galactic Nuclei (AGNs)

As we've seen, the central region of the M87 galaxy is considerably more hectic than the bulge of our own Milky Way. For this reason astronomers call M87 an *active* galaxy and its central region an Active Galactic Nucleus (AGN). The Milky Way is not considered an active galaxy.

It's all down to how much a galaxy's supermassive black hole is eating. If a lot of material is being sucked into the black hole it forms an accretion disc – effectively a large, flat rotating queue of material waiting to go in. As gas and dust spiral inwards quicker and quicker, friction rockets up the temperature and the super-heated material glows with high-energy ultraviolet light and X-rays. The centre of an active galaxy usually emits more energy than the rest of the galaxy combined. Some AGNs are so powerful that they would outshine over a thousand galaxies like the Milky Way.

AGNs can flare up, too – the amount of energy they're emitting spikes over a short period of time. It's thought this is because the supermassive black hole is gorging on a particularly big meal. Astronomers can tell the size of the meal from the duration of the spike. A flare-up lasting a week is probably caused by a cloud a light-week across (1/52nd of a light year).

In approximately one in ten active galaxies, interactions between the accretion disc and the black hole's magnetic field marshal some of the material into symmetrical jets that erupt at right angles to the disc. This is what's happening in M87. However, the jets are not escaping from within the black hole itself – that is still impossible. They are surging away from the accretion disc, which sits just outside the black hole's event horizon.

Quasars and blazars

The most powerful AGNs can be seen for huge distances across the universe. At first glance they appeared to be stars, but measurements of their distance revealed they are often billions of light years away. No ordinary star is bright enough to be seen from that far away, so they were dubbed 'quasi-stellar objects'. This was later shortened to just quasar.

What astronomers call an AGN depends on the angle we're seeing it from. If we happen to be looking straight down one of the jets then it is called a blazar instead. As these jets are very narrow, blazars are very compact objects. They are also highly variable, as the strength of AGN jets depends on how much gas the central black hole is consuming.

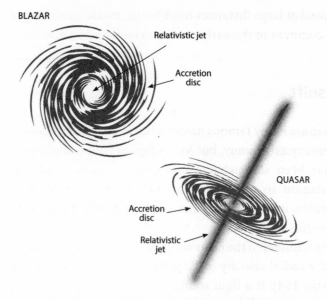

BLAZAR

Relativistic jet

Accretion disc

QUASAR

Accretion disc

Relativistic jet

Astronomers call an AGN either a quasar or a blazer
depending on the angle we see it from.

Looking at far-away objects such as quasars and blazars means looking back into the past. Imagine you receive a postcard from a friend telling you about their holiday. When you read it, you're not finding out about what they're doing at that moment – you're discovering what they were up to when they wrote it several days earlier. It took time for the message to reach you, so the postcard will always bring you news of the past, not the present. It is the same with light and space.

When we see an object a billion light years away, that light has taken a billion years to reach Earth. So we're seeing an image of the universe a billion years in the past. Astronomers have discovered that most quasars and blazars are found at large distances from Earth, meaning they were more common in the early universe compared to now.

Redshift

There are many famous names associated with twentieth-century astronomy, but Vesto Slipher isn't one of them. He has been somewhat overlooked by history, but his contribution to our understanding of the universe was invaluable: in 1912 he became the first person to measure the redshift of a galaxy.

We encountered the ideas of redshift and blueshift earlier, with the radial velocity method for discovering exoplanets (see page 154). If a light source is moving away from you, its light waves will be stretched out and its spectral lines – that black barcode pattern – will shift towards the red end of the colour spectrum. The lines of an approaching object are shifted towards the blue end. The more they shift, the faster the object is moving.

Slipher was the first astronomer to accurately analyse the spectra of galaxies and find these shifts. By 1921 he had examined forty-one galaxies in total, discovering that Andromeda and three others are moving towards us (their spectra are blueshifted). However, the majority of his galaxies were redshifted – they are running away from the Milky Way.

Today we know of around a hundred blueshifted galaxies, but hundreds of billions of redshifted ones. That means almost every galaxy in the universe is receding from the Milky Way.

THE HUBBLE ULTRA DEEP FIELD

The Hubble Space Telescope (HST) has completely changed the way we understand the universe.

One of its most famous photographs is the Hubble Deep Field. Between 18 and 28 December 1995, astronomers used the HST to peer relentlessly at an area of sky the size of a grain of sand held at arm's length. The photograph was full of three thousand specks, smudges and spots – some of the most distant galaxies ever discovered. They're so far away that most no longer exist. They've disappeared in the 13 billion plus years it has taken for their light to reach us.

Between 2003 and 2004, astronomers took a similar photograph called the Hubble Ultra Deep Field. Estimates based on it suggest that the observable universe contains 2 trillion galaxies. Each has hundreds of billions of stars, meaning there are more stars in the universe than there have been heartbeats in all of human history. One heartbeat per second for every *Homo sapiens* that has ever existed is still a thousand times lower than the number of stars out there.

Hubble's Law

The name most associated with redshifted galaxies is not Slipher, but that of his fellow American astronomer Edwin Hubble. Hubble measured distances to galaxies using the Cepheid variable technique pioneered by Henrietta Swan Leavitt (see page 182) and compared them to Slipher's data on the galaxies' redshifts. He found a very simple pattern: the further away a galaxy, the more it is redshifted. More distant galaxies appear to be moving away from us faster than those close by. Hubble published his work in 1931.

This rule has become known as Hubble's Law (although Belgian priest and astronomer Georges Lemaître published a similar idea in 1929). A number called Hubble's constant tells us how fast the galaxies appear to be moving. The modern value for Hubble's constant – given the symbol H_0 – is approximately 21 kilometres per second per million light years. If Galaxy A is a million light years further away from us than Galaxy B, it will recede an extra 21 kilometres every second.

Thanks to Hubble's Law, redshift has become a great way to measure distances in space. All you have to do is analyse the spectrum of a galaxy to find its redshift, before using Hubble's Law to convert that into how far away it is. The most distant object currently known – the one with the highest redshift – is GN-z11 some 13.4 billion light years away.

The expanding universe

Hubble's Law is a very simple premise: the further away a galaxy, the faster it appears to be moving away from us. Yet this seemingly innocuous idea has incredibly profound consequences. It means our universe is expanding.

At first glance it might not be immediately obvious why Hubble's Law implies that we live in an expanding universe. It helps to imagine some dough filled with raisins that you're about to put in the oven to bake. Let's say that the dough will expand to twice its original size in an hour. Put yourself in the place of one of the raisins and think about what you'd see. A raisin that was initially one centimetre away from you would end up two centimetres away. One that started two centimetres away would now be four. The nearer raisin will have appeared to move one centimetre in an hour, the further one two centimetres in the same time. More distant raisins appear to move away faster.

You could even say, 'raisins in an expanding dough appear to move one centimetre per hour for each centimetre of original separation'. That's exactly what Hubble's constant says: galaxies move 21 kilometres per second for every million light years of original separation. Just as the dough is expanding, so is the universe.

Galaxies aren't receding from us because they are moving away through space. After all, the raisins don't move through the dough. Instead, the gap between galaxies is stretching out as the space between them expands. The bigger the distance between us and a far-off galaxy, the more space there is to expand and the faster they'll appear to be carried away from us.

CHAPTER 6

The Universe

The Big Bang

Origins of the idea

Hubble showed us that the universe is expanding. A universe expanding today was smaller yesterday, so it is natural to think that it must have been very small in the distant past. This tallied well with earlier work by Alexander Friedmann and Georges Lemaître in the 1920s. They'd used the equations of Einstein's General Theory of Relativity to argue that the universe has expanded over time from an initial compact state.

We can use the rate at which the universe is getting bigger – Hubble's constant – to work backwards and calculate when the expansion started. The modern answer we get is 13.8 billion years ago. Watch the expansion in reverse and you would see everything getting closer and closer together. If you follow general relativity to the letter, all of space(time) ends up concentrated in a singularity – the same infinitely small, infinitely dense point it predicts sits at the centre of a black hole. The concepts of both space and time break down at a singularity.

Together these clues suggest that time and space began around 13.8 billion years ago when an incredibly small, hot point exploded outwards. Astronomers call this event the Big Bang. The universe it created has been expanding and cooling ever since.

The steady state model

The term 'Big Bang' was coined by English astronomer Fred Hoyle during a BBC radio interview in 1949. He was a major critic of the Big Bang, favouring instead the steady state model – the idea that the universe has been around for ever, pretty much in its present form. In direct contrast to the Big Bang, time and space have no beginning or end in a steady state universe. The theory was devised in 1948 by Hoyle, Hermann Bondi and Thomas Gold.

They sought an alternative picture because the Big Bang theory of the 1940s had a major issue: it said that the universe was younger than the Earth. Astronomers had grossly overestimated Hubble's constant – the measure of how fast the universe is expanding – because they couldn't accurately measure distances to galaxies. Thinking that the universe was expanding much more quickly than it was, they had massively underestimated its age. Hubble's original value was just 2 billion years. Geologists had already found rocks that were 3 billion years old.

The steady state model accounts for the observed expansion of the universe by saying that new matter is created to fill the gaps as space expands. That way, the overall density of the universe remains steady over time. It would mean new stars and galaxies eventually pop up alongside much older ones. In a universe that's steady, neighbouring stars and galaxies should be a mix of ages.

So, in the 1940s, as with many times in the history of science, there was a stand-off between two rival theories. The only way to proceed was for both theories to make predictions about what the universe should look like if they are right. Go out and find what you said you'd find and the spoils are yours to claim.

Nucleosynthesis

A steady state model doesn't have to explain how the universe came to look the way it does. It has always been here in its current form. The trouble with a Big Bang is that you're not only saying there was a beginning to space and time, but also that the universe was fundamentally different at the start to how it is now. If the Big Bang theory is to be believed, you need to explain how we ended up with a large universe full of stars and galaxies from an initial tiny, hot point.

If today's universe was once smaller than an atom, temperatures would have been exceedingly high – 10 billion degrees Celsius just one second after the Big Bang. Astronomers can use what we know about particle physics to say what would have happened under such extreme conditions. That's what particle accelerators such as the Large Hadron Collider are doing – recreating the environment immediately after the Big Bang.

Initially, the baby universe was only filled with energy. But, in that first second, temperatures were high enough for some of that energy to turn into matter. Protons, neutrons and electrons form – the building blocks of atoms. However, after just one second of expansion, the universe has cooled a little and no new particles can form in this way.

Some of the protons and neutrons then stick together to make particles called deuterons (a form of hydrogen nucleus). At three minutes old, the universe is hot enough for nuclear fusion to take place, but cool enough for the resulting particles not to get blasted apart. Some of the deuterons and protons fuse together to make the nuclei of helium atoms – the same process that turns hydrogen into helium in the centre of the sun (see page 51). Astronomers call this *nucleosynthesis*.

However, by the time the universe is twenty minutes old, it has cooled further and this fusion stops. Calculations suggest a quarter of the universe's hydrogen would have been turned into helium during that seventeen-minute burst of fusion.

This forms a fundamental prediction of the Big Bang theory. Once fusion stopped, there was no new way to change what the universe was made of. At least not until stars came along millions of years later and made a smattering of heavier elements. So the cosmos of today should still be largely 75 per cent hydrogen and 25 per cent helium. That's exactly what astronomers find when they look at the modern universe – a resounding tick for the Big Bang theory.

Where is all the antimatter?

The process that turns energy into particles is called pair production. As its name suggests, it always creates two particles – one matter, one antimatter. An antimatter particle is a mirror image of a normal particle. It has all the same properties, but the opposite electric charge. The antiparticle of the negatively charged electron is the positron, for example.

Pair production can create a particle–antiparticle pair so long as the energy is high enough to cover the masses of both (according to Einstein's famous equation $E=mc^2$). That's why the Big Bang theory has pair production stopping when the universe was just a second old. While still exceptionally hot, it had cooled sufficiently that the energy available couldn't cover the masses of any new particle–antiparticle pairs.

The opposite of pair production is annihilation, where a particle and its antiparticle meet and turn back into energy. As pair production should create equal quantities of matter and antimatter, in the 13.8 billion years since the Big Bang all matter should have annihilated with antimatter, leaving a universe once more filled with energy alone.

But that hasn't happened. There's loads of matter in the universe – stars, planets, people. Astronomers believe that for every billion antimatter particles originally created, a billion and one matter particles appeared. All the antimatter later annihilated with nearly all the matter. Everything you see around you is made from the tiny surplus of matter particles that didn't get annihilated. Why the universe had this slight bias for matter over antimatter is one of the biggest unanswered questions in physics.

'Re'combination

According to the Big Bang theory, fusion stopped after the universe had turned 25 per cent of its hydrogen into helium. It was just twenty minutes old at this point. But then nothing much happened for a very long time: 380,000 years. The universe was a sea of energy, electrons, protons (hydrogen nuclei) and helium nuclei that continued to expand and cool.

As we saw in Chapter 5, looking at objects in the distant universe is the same as looking back in time. But our view is blocked if we try to look back to distances equivalent to the universe's first 380,000 years. At that time the sea of particles was so dense that no light could escape. It's like trying to look through fog.

However, according to the Big Bang theory, the universe eventually expanded and cooled sufficiently that the protons and helium nuclei could grab hold of passing electrons to form atoms for the first time. This would have freed up considerable space and suddenly allowed light to surge outwards. Physicists call this event *recombination*. However, that's a terrible name for it because the electrons and nuclei had never been combined before.

Nevertheless, if the Big Bang really happened, the light released at the point of recombination should have flooded the universe. Over the last 13.8 billion years it will have a lost a lot of its energy, but it should still be there. This relic radiation is a key prediction of the Big Bang theory as it wouldn't exist in a steady state universe. Finding out whether or not it's there was crucial in deciding between the two.

The Cosmic Microwave Background

In 1964, American astronomers Arno Penzias and Robert Wilson were working with the Holmdel Horn Antenna in New Jersey. It had been built to pick up radio waves reflecting back from some of the first communication satellites to be launched into space. These signals were incredibly faint and so Penzias and Wilson were calibrating the antenna to remove any louder background noise including local radio broadcasts.

Yet, despite removing all the signals they could think of, there was still a quiet hum being picked up by the antenna. It was coming from all over the sky and was there 24/7. At first they thought it could be due to droppings from the pigeons roosting in the horn. They called them a 'white dielectric material'. The pigeons were evicted and their handiwork meticulously cleaned, but the noise remained.

Meanwhile, just down the road at Princeton University, a team led by Robert Dicke was searching for the relic radiation predicted to be left over from recombination 380,000 years after the Big Bang. When Dicke heard of the hum detected by Penzias and Wilson he famously said: 'Boys, we've been scooped.' We now call this radiation the Cosmic Microwave Background (CMB). It was discovered completely by accident, but it was a blow from which the steady state model would never recover. The CMB is cast-iron proof the universe began as a small, hot point.

By the time of the CMB's release, expansion had cooled the universe to around 3,000K (2,727 degrees Celsius). That's similar to the surface temperature of a red dwarf star and so the original light released by recombination would have been reddish. However, the continued expansion of the universe over more than 13 billion years has stretched out this light to wavelengths well below human vision. That's why today we pick it up in the microwave and radio parts of the spectrum. Its temperature is now just 2.7K (-270 degrees Celsius).

You don't need a massive horn antenna to pick up the afterglow of the Big Bang. On old analogue TVs you get a black and white hiss when you tune between channels. Similarly, you get a crackling between stations on analogue radios. One per cent of that interference is coming from the CMB. You're picking up the oldest light in the universe, the echo of the Big Bang, shifted to lower frequencies by its expansion.

The CMB is cast-iron proof our universe had a small, hot beginning.

Quasars

The year before the discovery of the Cosmic Microwave Background, Maarten Schmidt discovered the first quasar. These objects are the extremely bright cores of galaxies (see page 191). Since then, astronomers have found over 200,000 quasars. Almost all of them appear to be in the distant universe.

If the early universe contained lots of quasars, but our local (modern) universe doesn't, that suggests the cosmos has evolved over time. It cannot be in a steady state. Nor do we find any stars older than 13.8 billion years – the time that the Big Bang is said to have happened. Quasars make up one of the Four Pillars of the Big Bang theory.

The Four Pillars of the Big Bang theory are:

- The expansion of the universe
- Nucleosynthesis (75 per cent hydrogen/25 per cent helium)
- The Cosmic Microwave Background
- Quasar distribution

WHERE IS THE CENTRE OF THE UNIVERSE?

This is a really common question. People often assume *we* must be at the centre because we see galaxies moving away in all directions. But people in every galaxy would say the same. On page 196 we compared galaxies to raisins in expanding dough. Put yourself in the place of any of the raisins and you'll see all the others moving away. They can't, then, all be at the centre.

Astronomers are often asked to point to the place where the Big Bang happened, but that's not possible. Perhaps because the Big Bang is often compared to an explosion, people imagine a bomb going off. If a bomb explodes in a room, the debris can be used to say where in the room it detonated. The difference is that the Big Bang created space. Imagine that an exploding bomb created a room and then ask where in the room it exploded.

Pick any point in the universe and imagine where it was at the Big Bang. It was part of the explosion. This is why astronomers say that the Big Bang happened everywhere at once.

Problems with the Big Bang

The Big Bang is far and away the best theory we have to explain where the universe came from. All evidence points to a small, hot beginning. However, there are some niggling problems that still need addressing.

How can something come from nothing?

In the original version of the Big Bang theory, the universe begins as a singularity – the infinitely small, infinitely dense point predicted by Einstein's Theory of General Relativity. Infinitely small means it has no size at all. It was literally nothing. But how can something come from nothing?

Except that singularities probably aren't a real feature of the universe. They're more a glaring, neon signpost that says we don't quite understand the physics correctly. As we saw in Chapter 4, physicists are trying to combine Einstein's landmark theory with quantum physics to create a more complete Theory of Everything (see page 147).

We already know that something *can* come from nothing in the quantum world. Even in a perfect vacuum, energy is turned into pairs of particles that swiftly disappear again. Physicists call them *virtual* particles. The same particles involved in Hawking radiation from black holes (see page 146). A Theory of Everything could show us that Einstein's fabric of space-time is not continuous but made out of a series of bubbles. If so, these bubbles could come and go just like virtual particles.

It might then be possible that our universe didn't come from nothing, but arose from a tiny bubble in space-time.

Almost a singularity, but not quite. However, we'd need a reason for why our bubble expanded and didn't just disappear again. There's nothing in the original version of the Big Bang theory that can explain that.

What happened before the Big Bang?

This question is the sibling of *how can something come from nothing?* The original version of the Big Bang says time started with the explosion of a singularity. Just as there's nothing north of the North Pole, there's nothing before the earliest point in time.

That answer is not satisfactory to most people, especially when you consider the usual rules of cause and effect. Say you drop this book. It hitting the floor (the effect) will happen after you've let it go (the cause). We're so familiar with this idea that if you only saw the book hitting the floor you'd be right to assume that sometime earlier someone had dropped it.

If the Big Bang was the effect, what was the cause? If the effect created time, how can there even be a prior cause? With the original Big Bang model it is not possible to talk about a time before the Big Bang.

Magnetic monopoles

According to the original Big Bang theory, the early universe would have been hot enough to create magnetic monopoles – hypothetical particles with only one magnetic pole. Yet physicists have never seen a single magnetic monopole anywhere in the universe.

Temperature variations in the Cosmic Microwave Background

When recombination released the light we now see as the Cosmic Microwave Background, the temperature of the universe was around 3,000K (2,727 degrees Celsius). But today the radiation we pick up from the CMB corresponds to a temperature of just 2.7K because the universe has expanded significantly (see page 203).

Astronomers have exquisitely mapped out the Cosmic Microwave Background using satellites such as *WMAP* and *Planck*, detecting tiny temperature deviations equal to just one part in a million. Some parts of the CMB are ever so slightly hotter or colder than the rest. That means some parts of the early universe were slightly hotter or colder when the CMB was released.

This can be explained if matter in the early universe wasn't evenly distributed. Slightly denser regions would have been hotter and sparser areas cooler. That also fits with the modern structure of the universe, where huge superclusters of galaxies are surrounded by vast supervoids. The sparse regions were stretched out by expansion to form the voids and the gravity of the denser regions drew in additional material to form clusters. However, the original Big Bang model has no explanation for the origin of tiny variations in the distribution of matter in the early universe.

The horizon problem

The tiny temperature fluctuations aside, the Cosmic Microwave Background is incredibly smooth. How is it that the background temperature is the same throughout the entire observable universe? If you open a window on a winter's day, heat will flow out until the inside becomes as cold as outside. A physicist would say the two places eventually reach *thermal equilibrium*. But it takes time to get there. Like everything else in the universe, the maximum speed at which anything can be exchanged between two locations is the speed of light. That's never going to be a problem in your house, but it is in space.

Let's look at a patch of sky 10 billion light years away in one direction and then another patch the same distance away in the exact opposite direction. That means there's a distance of 20 billion light years between them. The universe is only 13.8 billion years old, so how have those two regions of space had time to reach thermal equilibrium?

You might say that they were closer together in the past, but they've never been close enough. The Big Bang theory tells us how fast the universe has been expanding since the beginning. To get as far apart as they are today, these two regions of space can never have been near enough to reach thermal equilibrium. Light has never been able to travel from one to the other – they've always been beyond each other's horizon. This *horizon problem* is one of the biggest issues with the original version of the Big Bang.

The flatness problem

The Earth's surface is curved, but for that curvature to become apparent you need to see or travel a reasonable distance across it. Imagine that you are stuck in one spot and can only ever see an area within ten metres of you. You'd think the Earth was flat, even though it isn't.

That situation is similar to our experience of the universe. We are currently restricted to the solar system and rely on light bringing us details about what's further afield. However, we can only see objects if that light has had time to reach us. The universe expanded so fast in the beginning that there are parts of it we'll never see. So there's a distinction to be made between the universe (all that exists) and the *observable* universe (the part we are able to see).

Measurements of the observable universe suggest that the space within it is flat – it has no discernible overall curvature. There are two possible explanations for this. First, the expanding universe has stretched space out so much that the small piece of it visible to us *appears* flat, even though the wider universe is curved. This is the same as your ten-metre area on Earth looking flat when the planet's surface really curves. However, according to the original version of the Big Bang, the universe hasn't stretched enough for that to happen. So either the Big Bang isn't the whole story, or the entire universe – the bit we can see and the bit we can't – is flat. Astronomers have calculated that the odds of that happening are about one in a hundred trillion trillion trillion trillion trillion.

The fine-tuning problem

The apparent flatness of our universe isn't the only thing about it that is incredibly unlikely. Imagine that our universe has some giant control panel with a series of knobs, dials and buttons. Each of them governs one aspect of the universe. It could be the speed of light, the mass of an electron or the strength of gravity. If you change any one of these settings – even by a few per cent – our universe would have turned out very differently.

Let's take gravity. If it were stronger, material would get crushed together more intensely in the heart of stars. A star would rocket through fusion a lot faster and stars would last months or years, not the billions of years they do now. Life on Earth wouldn't have had a chance to arise under such conditions. Change the dials enough and stars don't form at all. If gravity was considerably stronger, it could have overturned the universe's initial expansion and caused everything to collapse back in on itself in a 'Big Crunch' before the first stars ever ignited.

If all of these settings are random, and could have taken on a wide range of values, then how come all the dials are in the perfect place to give rise to a universe full of stars, planets and people? Most other settings would yield an empty universe or none at all. There are several answers to this *fine-tuning* problem. It could just be dumb luck – unlikely things do sometimes happen. The second is that some omnipotent Creator deliberately designed it that way. Neither of these explanations is satisfactory because they're not testable.

However, a third explanation – an idea called *inflation* – has the potential to explain not only the fine-tuning problem, but all the other problems with the Big Bang to boot.

Inflation

Fixing the Big Bang's problems

By the late 1970s, many of these problems with the Big Bang had become apparent. It was clear that some sort of Big Bang happened because the evidence from the Cosmic Microwave Background, nucleosynthesis and quasars was too hard to ignore. However, something had to give.

Starting in 1979, and continuing into the early 1980s, physicists Alan Guth, Andre Linde and Paul Steinhardt devised a way to alter the Big Bang idea ever so slightly while maintaining all of its strengths. Their idea is called

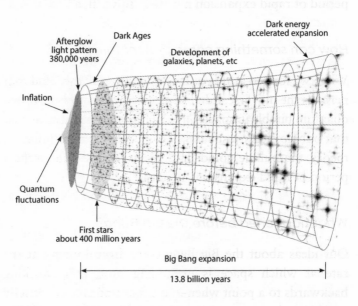

Our best picture of the universe's history, from the initial period of inflation to the dark-energy dominated era of today.

inflation and its premise is incredibly simple: in its earliest moments the universe underwent a period of expansion far more rapid than anything that would come later. Think of it as the original expansion envisaged by Hubble, but on steroids. In the first trillionth of a trillionth of a trillionth of a second, the universe went from much smaller than an atom to about the size of a grapefruit. It may not sound like much, but that's a scale factor of one with seventy-eight zeroes after it. If you made a red blood cell that many times bigger you would have something a trillion trillion trillion times wider than the observable universe.

If we take each of the Big Bang's problems in turn, we can look at how the addition of an early inflationary period of rapid expansion may help solve them.

How can something come from nothing?

When we looked at this question on page 206 we said that perhaps the universe didn't come from nothing, but from a quantum bubble in space-time. Yet we needed a reason why that bubble didn't just vanish again. According to inflation theory, the bubble could have survived if it underwent a period of rapid expansion similar to inflation.

What happened before the Big Bang?

Our ideas about the Big Bang come from looking at the rate at which space is expanding now, then working backwards to a point when that expansion started. Strictly speaking, that expansion – the bit that obeys Hubble's Law – only began after inflation had ended. So inflation is what happened before the Big Bang. Many theorists argue you don't need a singularity before inflation, particularly if there really is a Theory of Everything. Whatever was there before an area of it inflated to form our universe could well have been there for ever.

Magnetic monopoles

A period of inflation would have rocketed any magnetic monopoles much further apart than the original Big Bang picture suggests. They'd be so far spread out by now that it's no surprise we've never encountered one.

Temperature variations in the CMB

We know that on the smallest scales there are always virtual particles popping in and out of existence (see page 146). These quantum fluctuations cause temporary changes in the amount of energy at any given point in space. They would have been magnified to astronomical scales during inflation, leading to areas of the new universe with more or less energy than the average.

That explains why the Cosmic Microwave Background has small temperature variations. Physicists get a great match when they compare the expected size of the inflated quantum fluctuations to the size of the temperature variations in the CMB. As we've seen, those variations became the seeds around which superclusters and supervoids would later form (see page 208). So inflation also explains why the structure of the modern universe looks the way it does.

The horizon problem

Inflation caused the universe to expand a lot faster in the beginning than the original Big Bang theory suggested. That means two areas of space could have started out a lot closer together and still be as far apart as they are now. If all points in space were a lot closer prior to inflation, they could have reached thermal equilibrium before being rocketed apart.

The flatness problem

One solution to the flatness problem is that space was stretched so much in the beginning that the observable universe appears flat to us, even though the wider universe may have some curvature (much like the Earth would appear flat from a small area of it).

Except we said the Big Bang alone couldn't have stretched the universe enough to make that true (see page 210). However, it makes more sense if there was a period of inflation causing a greater amount of expansion than we'd previously considered. Inflation would have smoothed out any curvature in the observable universe.

Fine-tuning and eternal inflation

That just leaves the fine-tuning problem and for that we need to consider the idea of *eternal inflation*.

Inflation offers an attractive solution to the major issues with the Big Bang. However, if you're going to claim that this period of rapid expansion really happened, you need to offer an explanation for why it inflated and how it transitioned into the universe described by the Big Bang.

To achieve this, inflation theorists invoke the presence of an inflationary field. In physics, a field is an area over which a force operates. The Earth has a gravitational field, for example. Its strength varies over the Earth's surface – it is stronger over mountains and weaker above valleys. Physicists believe the inflationary field varies, too. Inflation happens in areas where it is strong and stops in areas where it is weak. When inflation stops, the energy locked up in the inflationary field is converted into matter and radiation: Big Bang.

However, the only way theorists can get the energy of the inflationary field to transition neatly into something resembling the Big Bang is if it doesn't all transition at once. Each time part of it transitions you get another Big Bang, creating a new isolated region of space while inflation continues elsewhere. This is *eternal inflation* and its consequences are profound.

Multiple Big Bangs mean multiple universes. According to inflation theory, there should be a near infinite – maybe even truly infinite – number of them. The laws of physics, masses of particles and strength of forces will be different in each one depending on the exact way it transitioned from inflationary field to Big Bang. That's the equivalent of the knobs, dials and buttons on each universe's control panel having slightly different settings (see page 210).

If you think that your universe is the only one, then of course you're going to find it puzzling that its control panel is perfectly adjusted for your existence. You might even have illusions of a Creator. But if you realize your universe is just one of many, where are you going to find yourself?

You're not going to be in a universe where the settings mean you can't exist, where stars and planets cannot form. You can *only* be where the dials are in the right place. Eternal inflation solves the fine-tuning problem by saying there is an infinite inflationary multiverse out there and the full range of settings plays out across it. The dials have to be 'right' somewhere and you can't be anywhere else.

The multiverse

The idea of a multiverse takes some getting used to. It is a kaleidoscope of possibilities where everything that can happen does happen, somewhere. If the multiverse is infinite then every possibility happens an infinite number of times.

To see why this is the case, imagine you roll six dice. What are the chances of getting 1, 2, 3, 4, 5, 6? The answer is 1.5 per cent. So, on average, this pattern should turn up three times for every two hundred rolls. The more times you roll, the more times you'll see the same pattern.

It is exactly the same with the multiverse. Every time the inflationary field transitions into a Big Bang is another roll of the dice. Roll the dice enough times and you're likely to see the same pattern (universe) repeating. Roll an infinite number of times and you're guaranteed to.

Journey across an infinite multiverse and you'll eventually come across another universe where all its atoms are arranged in an identical pattern to this one. Every single atom. That includes the atoms in my fingers that typed these words because I was inspired to pursue a career in astronomy by twinkling atoms in the night sky when I was a kid, down to the atoms in your eyes that are receiving light from this page. In another part of the multiverse, you're doing the very same thing – that complete scenario being repeated.

What does it say about the choices you make if there are a million other yous in a million other universes making exactly the same choices? And a million other nearly-yous making entirely different choices? There are universes out there where you're the President of the United States and others where Washington is still governed from England. Loved ones who've died in this universe are alive and well in others. In some you have the head of a chicken or the pouch of a kangaroo. In an infinite multiverse, every possible configuration of atoms is guaranteed an infinite number of times.

Evidence for inflation

Multiple universes seem a natural consequence of eternal inflation, which in turn seems to help explain the features of our universe and improve upon the Big Bang. However, we currently have *zero* evidence that inflation, eternal or otherwise, really took place. In fact, Paul Steinhardt, one of the fathers of the theory, has turned his back on it. He has since been an outspoken critic of the multiverse.

However, many other researchers believe it is possible to find evidence of inflation. In fact, back in 2014, a team of scientists made headlines around the world with claims they'd found the equivalent of a smoking gun. The result came from the BICEP2 experiment located at Antarctica's Amundsen–Scott South Pole Station. The scientists had used it to look again at the Cosmic Microwave Background.

It is believed that an expansion as rapid as inflation would have sent gravitational waves rippling out through the infant universe. One day we might detect these primordial gravitational waves, but they are now very small over 13 billion years later, too small for our current gravitational wave detectors. However, the Cosmic Microwave Background could come to our rescue as it offers us a snapshot of what the universe was like at just 380,000 years old. If the universe were forty years old, the CMB is an image of it as a ten-hour-old baby. Any primordial gravitational waves passing through when the CMB was released should have left telltale twists in its light. In March 2014, the BICEP2 team told the world they'd found those twists.

Except most astronomers now agree they didn't. Doubts were expressed fairly rapidly, with the team behind the *Planck* satellite arguing that the same effect could be generated as the CMB light passed through dust in our own Milky Way much later. So, for now, astronomers are still on the hunt for the first evidence of inflation.

The BICEP2 debacle happened eighteen months before gravitational waves from colliding black holes were discovered for the first time by the LIGO experiment in September 2015 (see page 141). LIGO isn't sensitive enough to pick up primordial gravitational waves, but

THE UNIVERSE IN BITE-SIZED CHUNKS

now gravitational waves have finally been confirmed there's likely to be a rush to build bigger and better detectors. One day, these machines could show us the way to the multiverse.

THE CMB COLD SPOT

A lack of primordial gravitational waves hasn't stopped some scientists claiming they've already found evidence of other universes. It all hinges on an unusually cold area in the CMB.

First spotted by the *WMAP* satellite in 2004, it was picked up again by the *Planck* satellite in 2013. It's 140-millionths of a degree colder than the average CMB temperature of 2.7K, far greater than the normal temperature variations and too big to have been caused by a quantum fluctuation magnified by inflation.

Perhaps light from that area of the CMB has journeyed through a particularly large supervoid – an area with far fewer galaxies than the universal average. Losing energy on its way through, we'd see it as colder. Except an extensive survey of 7,000 galaxies in 2017 revealed no such void.

Other astronomers claim that the cold spot is evidence of the effect of another universe upon our own. During eternal inflation we could have bumped into a neighbouring bubble universe, leaving a 'bruise' in the CMB. It remains a highly controversial idea.

The frontiers of the universe

The edge of the universe

It is still unclear whether our universe is the only one or whether it is part of some infinite inflationary multiverse. If it's just us, does the universe ever stop? If there are many, where does the next one begin? Is there an edge to our universe?

There's certainly an edge to what we can see. The light of the Cosmic Microwave Background comes from the boundary of the observable universe – it was the first light capable of escaping from the dense fog of particles in the early universe. It marks our cosmic horizon, a bit like your horizon on Earth. Stand outside your house and you can only see so far. Yet you know the Earth doesn't stop at your horizon. Nor do astronomers think the universe stops at our cosmic horizon.

Most models of a solo universe have it continuing for ever, an infinitely big cosmos without edge or boundary. People often ask: *What is the universe expanding into?* But if our universe is the only one, then it contains, by definition, everything that exists. If there were somehow something beyond the universe, then it would exist and therefore be part of the universe. We know that the universe isn't expanding in the sense that galaxies are surging outwards *through* space into some previously unoccupied area. Instead, it's the space between galaxies that is stretching (see page 196).

If our universe is part of a vast multiverse then all the individual universes would be part of some wider structure. Laura Mersini-Houghton – the physicist who believes the cold spot in the Cosmic Microwave Background is a bruise from another universe – has calculated how far away the next universe would be now. Her answer: at least a thousand times further away than our current cosmic horizon.

The fate of the universe

What lies ahead for the cosmos? The answer depends on how much stuff it has in it.

Our universe has been expanding since the Big Bang, galaxies flying apart as the space between them stretches. But the galaxies have a gravitational attraction towards each other, too. If there's enough matter and dark matter in the universe, their collective gravity will eventually overturn the expansion and start drawing galaxies closer together again. The universe will collapse in a 'Big Crunch'. If there's not enough stuff in the universe, the expansion will continue, gradually slowing but never stopping. The third possibility is that there's just enough mass in the universe to bring the expansion to a halt, but not enough to trigger a collapse.

All three scenarios have one thing in common: the expansion of the universe should currently be slowing down. In the mid-1990s, two teams of astronomers were working on projects to determine exactly how the rate of the universe's expansion has changed over time.

As we saw in Chapter 5, looking at far-away objects is the same as looking back in time. Light is like a postcard bringing us information from the past (see page 192). The further away a galaxy, the further back in time it represents. By measuring the speed of distant galaxies we can see how fast the universe was expanding long ago and compare that to the speed of expansion now. If the universe is slowing down, the expansion should have been quicker in the past.

However, we need to accurately measure how far away distant galaxies are if we want to know what point in the universe's history they represent. The normal distance-determining methods of parallax and Cepheid variables don't work for such far-flung places (see page 182). The two teams of astronomers needed a new, much brighter standard candle called a Type Ia supernova.

Type Ia supernovae

Our sun is an unusual loner. Most stars exist in pairs, like the twins suns of the planet Kepler-16b (see page 156). Imagine a situation where one of the duo dies and forms a white dwarf, just as our sun will do eventually. This dense core has a mighty gravitational pull and so it starts to rip gas away from its companion. Gorging on more and more material, the white dwarf gets heavier and heavier.

However, there is a limit to how much a white dwarf can consume. Known as the Chandrasekhar limit, it was calculated by Indian astrophysicist Subrahmanyan Chandrasekhar when he was just nineteen years old. In 1930, he took a boat from the Indian port of Madras to Genoa in Italy en route to Cambridge. During the three-week voyage he calculated that the mass of a white dwarf can never exceed the equivalent of 1.4 suns. As a white dwarf approaches this limit it becomes unstable and violently explodes as a calamitous supernova. Astronomers call these events Type Ia supernovas to distinguish them from the core collapse explosions at the end of a big star's life (a Type II supernova; see page 136). They are the perfect standard candles: not only are they exceptionally bright, meaning they can be seen halfway across the universe, but they always have a similar inherent brightness. Every time one detonates it does so with around 1.4 suns' worth of fuel. A fixed amount of fuel means a fixed brightness.

The Type Ia supernova known as SN 1994D exploding in the galaxy NGC 4526. Notice how a solitary explosion can shine as bright as an entire galaxy's hectic core.

To work out the distance to the galaxy in which the star exploded, all you have to do is compare how bright it appears in the sky to how bright it should be (its apparent magnitude to its absolute magnitude). The bigger the difference, the more the light has faded during a longer journey to the Earth.

In 1998, the two teams working on the expansion history of the universe published results based on measurements of Type Ia supernovas. To everyone's utter astonishment they found that the universe's expansion appears to be getting *faster*. In 2011, three of the scientists behind the discovery were awarded the Nobel Prize in Physics.

Dark energy

It seems that after the Big Bang the expansion rate of the universe initially slowed as expected, but about 6 billion years ago it suddenly started speeding up again. Nobody saw this coming and it remains baffling.

There appears to be something overpowering gravity and actively pushing galaxies apart. The influence of this mysterious entity must have been fairly negligible in the early universe, but grew in stature as the cosmos aged. Astronomers call this shadowy anti-gravity *dark energy*, in the same vein as the invisible dark matter thought to glue galaxies together (see page 171). Astronomers believe our universe is currently 68 per cent dark energy and 27 per cent dark matter. The atoms that make up ordinary matter – the same stuff you and I are made of – comprise just 5 per cent.

Dark energy is really a placeholder name because we know even less about it than we do about dark matter. The leading idea is something called vacuum energy and it's a concept we've encountered several times before. Empty space is never truly empty; there are always virtual particles popping in and out of existence (see page 146). The average amount of this vacuum energy in a given area of space is always the same. In the early universe, the gap between galaxies was small, so there wasn't a lot of vacuum energy around. However, the momentum of the Big Bang stretched the space between galaxies over time. More space equals more vacuum energy. Eventually there was so much space that the repulsive power of vacuum energy trumped the waning pull of gravity and the universe started accelerating.

This picture may sound neat and tidy, but it has gaping holes in it. According to quantum physics, the amount of energy in the vacuum should be 10^{120} times greater than we observe. That's one with 120 zeroes after it! If dark energy were just vacuum energy then the universe would have been ripped apart long ago. This is another example of a fundamental disagreement between quantum physics and general relativity. We may have to wait for a Theory of Everything to emerge before we have the tools to understand dark energy.

The Big Rip

If dark energy is accelerating the expansion of the universe then all the outcomes we looked at on page 222 are off the table. Instead, space will keep on stretching at an ever-faster pace. More space will mean more dark energy, leading to ever-quickening expansion. It's a vicious, runaway effect.

Eventually the space between stars will stretch so much that it overcomes dark matter's adhesive powers and galaxies will fly apart. The space between stars and their planets will eventually grow, too, and solar systems will buckle under the expansion.

Gravity is by the far the weakest force, so these gravitationally bound systems will go first. The electromagnetic force that holds electrons to the nuclei of atoms will fall next. The expansion of space between electrons and the nucleus will overcome this force and atoms will fly apart. Eventually, even the strong nuclear force that holds protons and neutrons together in the nucleus won't be able to stem the rising tide of dark energy.

The upshot is that everything in this universe will be torn to shreds. Astronomers call this the *Big Rip*. No galaxies, no stars, no planets, no atoms. A vast, empty sea of nothingness. Calculations suggest this universe of ours will die approximately 22 billion years from now.

Conclusion

'The sight of the stars always makes me dream.'
Vincent Van Gogh (1888)

Nothing lasts for ever. That we are here in the middle of this universe's existence to ponder its greatest mysteries is a privilege and one that we should cherish.

Humanity has been on a quite remarkable astronomical journey. At first we thought we were the centre of everything, the sun and stars bending to our will. Then our logic and reasoning relegated us to just another planet around just another star in just another galaxy in a small corner of a colossal universe. This universe could well be one in an infinite expanse of others where every conceivable drama plays out on every imaginable stage.

Such discoveries are enough of a reward, yet some people ask why we bother to explore space. The answer is that it's in our DNA. Our curiosity led us out of Africa and across the world, to the top of Mount Everest and to the bottom of the oceans. We've witnessed Earthrise on the moon and sunset on Mars, and seen to the edge of the observable universe. We have a drive to know what's out there and to push the boundaries of what's possible.

There is a high probability that people alive today will

see the first human mission to Mars, the first time we'll ever set foot on another planet. Today's schoolchildren are the Martian colonists of tomorrow, forging a new path for humanity across the solar system. In the decades ahead, our telescopes may well also reveal undeniable proof that we're not alone in the universe.

For those who say that curiosity alone is not enough to justify these endeavours, there is a more practical concern. While we remain a one-planet species all our eggs are in the same Earth-sized basket. Only by venturing out into space will we give ourselves the best chance of survival should a stray asteroid, lethal pandemic or nuclear war threaten our future.

We came from space, after all. The calcium in your bones and the iron in your blood were forged in the heart of dying stars and blasted across space by mighty supernova. By venturing into space we're only going back home, and our efforts in astronomy and space exploration are catapulting us towards a permanent human presence in space.

So until the Big Rip brings the curtain down on this universe, long may we continue to look up in awe and wonder.

Index

(page numbers in italics refer to photographs, diagrams and captions; those in bold refer to tables)